Franklin Hadley Cocks

**Energy Demand and
Climate Change**

Related Titles

Coley, D.

Energy and Climate Change

Creating a Sustainable Future

2008
ISBN: 978-0-470-85312-2

Paul, B.

Future Energy

How the New Oil Industry Will Change
People, Politics and Portfolios

2007
ISBN: 978-0-470-09642-0

Kruger, P.

Alternative Energy Resources

The Quest for Sustainable Energy

2006
ISBN: 978-0-471-77208-8

Olah, G. A., Goeppert, A., Prakash, G. K. S.

**Beyond Oil and Gas: The Methanol
Economy**

2006
ISBN: 978-3-527-31275-7

Synwoldt, C.

Mehr als Sonne, Wind und Wasser

Energie für eine neue Ära

2008
ISBN: 978-3-527-40829-0

Romm, J. J.

Der Wasserstoff-Boom

Wunsch und Wirklichkeit beim Wettlauf um den
Klimaschutz

2006
ISBN: 978-3-527-31570-3

Franklin Hadley Cocks

Energy Demand and Climate Change

Issues and Resolutions

WILEY-VCH

WILEY-VCH Verlag GmbH & Co. KGaA

The Author

Prof. Franklin Hadley Cocks
Duke University
Pratt School of Engineering
Durham NC 27708
USA

■ All books published by Wiley-VCH are carefully produced. Nevertheless, authors, editors, and publisher do not warrant the information contained in these books, including this book, to be free of errors. Readers are advised to keep in mind that statements, data, illustrations, procedural details or other items may inadvertently be inaccurate.

Library of Congress Card No.: applied for

British Library Cataloguing-in-Publication Data
A catalogue record for this book is available from the British Library.

Bibliographic information published by the Deutsche Nationalbibliothek
The Deutsche Nationalbibliothek lists this publication in the Deutsche Nationalbibliografie; detailed bibliographic data are available on the Internet at <http://dnb.d-nb.de>.

© 2009 WILEY-VCH Verlag GmbH & Co. KGaA, Weinheim

Typesetting SNP Best-set Typesetter Ltd., Hong Kong
Printing Strauss GmbH, Mörlenbach
Binding Litges & Dopf GmbH, Heppenheim

Printed in the Federal Republic of Germany
Printed on acid-free paper

ISBN: 978-3-527-32446-0

To the memory of my parents,
Ruth and Charles

Contents

Energy Demand and Climate Change: Issues and Resolutions. Franklin Hadley Cocks
© 2009 WILEY-VCH Verlag GmbH & Co. KGaA, Weinheim
ISBN: 978-3-527-32446-0

Acknowledgements

The author wishes to thank the following people (in alphabetical order), for their help far above and beyond the call of friendship and for their patience in reading and enhancing a panoply of subject matter: Frank Gayle, Ulrich Goesele, Charles Harman, George Hatsopoulos, George Hurley, See-Whan Kang, J.D. Klein, Ian Pollock, Bob Rose, Neal Simmons, Craig Tedmon, Steve Vogel, Seth Watkins, and David Wong.

Charles Harman, George Hurley, and Bob Rose deserve special thanks for helping me with this project almost from its very beginning. Special thanks also are due to Ulrich Goesele, Director of the Max Planck Institute for Microstructure Physics in Halle, Germany, who provided a place for me to work during my sabbatical leave from Duke University in the autumn of 1999, when this book had its genesis, and who has helped enormously in bringing it to publication. Linda Martinez, Head of the Vesic Library for Engineering, Mathematics, and Physics at Duke University, used her unparalleled information-retrieval expertise unstintingly to locate innumerable hard-to-find documents.

My sons, Josiah and Elijah, have provided steadfast encouragement, patient proofreading, and insightful perspectives. My wife, Pamela, has been my confidante, inspiration, spur, and friend *par excellence* through all the trials and tribulations in completing this book. To all of you I am deeply grateful. Any errors are mine and mine alone.

Energy Demand and Climate Change: Issues and Resolutions. Franklin Hadley Cocks
© 2009 WILEY-VCH Verlag GmbH & Co. KGaA, Weinheim
ISBN: 978-3-527-32446-0

Prologue

Global warming will pluck the strings of Nature's many instruments, but we may not like the melody they play. Much has been said about climate and the interdependence of civilization and energy. Numerous writings advocate particular aspects of the problems that climate change and energy shortages will cause. Some people take the position that there is no problem at all. The book now in your hands presents the facts – the scientific and engineering rules of the game – that govern the chess match now underway between humanity and nature, so that you may judge for yourself what is happening and the validity of the various positions being advocated. Science and engineering truths are independent of political viewpoint or vested interest.

A huge knowledge base envelops every facet of the energy and climate debate. The goal of this book is to pull together the fundamental facts of this ongoing saga and to present the near-term and long-run choices ahead of us and their consequences. The temperature of our planet has changed repeatedly in the past and is in the process of doing so again, with far-reaching and complex effects that will slowly unfold. Knowing what has happened in the past can help us to understand what is underway now. The fossil, nuclear, and renewable resources of our planet are a guide to planning what might be done, while there is still time.

Part I gives an overview of the human use of energy as it has evolved through the ages as well as the astronomical and atmospheric factors that have dominated our planet's climate. Earth's slow but inevitable orbital changes have an enormous and long-term influence on global climate, especially the periodic onset of ice ages. Humanity's ever-expanding consumption of energy has contributed greatly to the betterment of living standards, which depend critically on fossil fuels, whose supply is not infinite. Earth's nuclear fuel resources are large, but making use of them generates its own special problems.

Part II presents energy options that can be called into being with the technology that exists right now. Increased efficiency of energy usage and energy from renewable resources including wind, sunlight, and many others offer a variety of possibilities, each having different potentials and limits.

Part III discusses the energy and climate-changing possibilities that are only dreams now but might someday come to be. Thermonuclear fusion, breeding nuclear fuel, artificial changes in planetary albedo, magnetohydrodynamic

Energy Demand and Climate Change: Issues and Resolutions. Franklin Hadley Cocks
© 2009 WILEY-VCH Verlag GmbH & Co. KGaA, Weinheim
ISBN: 978-3-527-32446-0

electricity production, power from ocean thermal and salinity gradients, and other technologies are possible. Each of these also has both potentials and limits.

Part IV offers a glimpse of the devastating energy and climate possibilities that might envelop us if we just keep going along the way we are.

Our age is filled with problems and promises. The more people there are who understand the basic facts of the energy and climate events now underway and the options we have for dealing with them, the better chance there is for all of us to find a path that leads to a more abundant future for ourselves and our posterity. The choices we make now may determine whether or not our age marks the onset of Nature's checkmate move.

Part I Questions

Energy Demand and Climate Change: Issues and Resolutions. Franklin Hadley Cocks
© 2009 WILEY-VCH Verlag GmbH & Co. KGaA, Weinheim
ISBN: 978-3-527-32446-0

Introduction

Civilizations come and go, but why? There are many reasons, especially the proverbial four horsemen of the Apocalypse–war, famine, disease, and death. Battles have been won or lost. Droughts have desolated rich agricultural lands, and animals have been hunted to extinction. Epidemics have waxed and waned. In the 21st century our planet supports a larger population than ever before in its long history. The climate of the world has begun to change, and that unfolding event will affect everyone. Those four horsemen might begin to saddle up, armed now with nuclear weapons and virulent diseases. The number of mouths to be fed and the world's demand for energy grow larger with each passing day. The Earth is not infinite in extent, and neither are its resources. How is this to end?

Recent history shows that average birth rates may decline as living standards improve, and in many lands living conditions have been progressing. Before the advent of chemical means of birth control, better living conditions usually led to an increased rate of population growth, except under particular conditions. The French aristocracy in the 18th century, for example, made special attempts to limit any increase in their numbers in order to decrease difficulties associated with inheritance and the subdivision of estates. But as a general rule, increased prosperity can reduce population growth by making birth control and education available to more men and women, who have a greater expectation that their children will survive to adulthood. Overpopulation, posited in 1798 by the English demographer Thomas Malthus in his Essay on the Principle of Population, has been kept at bay by improved farming technology, genetic manipulation of crops, better education, birth control, greater prosperity, and the increased use of energy. In the year 2000 the world's population was 10 times higher than it was 300 years earlier. The population of the Earth has increased from 2.5 billion to more than 6 billion since 1950 alone, and average energy consumption per person more than doubled in that same period. The peril of runaway population growth might be eliminated if the world's economic output could increase sufficiently. Standards of living and energy consumption rise in unison. The energy from fossil fuels is the horse out in front pulling the world's economic wagon, but fossil fuels are not inexhaustible. Petroleum is especially limited in its total planetary supply. When there is demand for more oil than the Earth can readily yield, its cost will increase until supply and

Energy Demand and Climate Change: Issues and Resolutions. Franklin Hadley Cocks
© 2009 WILEY-VCH Verlag GmbH & Co. KGaA, Weinheim
ISBN: 978-3-527-32446-0

demand balance. Living standards may be depressed by the weight of higher energy costs.

The world economy is now interconnected as never before, and changing energy prices and economic crises spin rapidly around the globe. Coal remains abundant, but burning it releases carbon dioxide and a host of pollutants, including mercury, sulfur dioxide, radioactive isotopes, and many others. Natural gas is also abundant, and burning it produces much less carbon dioxide than burning coal, but its supply also has limits. Nuclear generation of energy produces radioactive waste, which must be stored indefinitely somehow, but has little effect on the atmosphere. Advances in engineering and science may yet meet ever-growing energy needs, if proper measures are adopted and steps taken in good time. Unlike the steady growth of population predicted by Thomas Malthus, primitive societies, especially those on isolated Pacific islands, sometimes suffered rapid declines in population as local resources were depleted. In modern societies, energy problems might develop rapidly, with accompanying disruptions of the social order, some of them violent. The current rate of growth of the world economy and extrapolated demographics based on increasing living standards indicate a topping out of world population at just over 10 billion souls. What if energy shortages cause the world economy to decline? Living standards would very likely degrade, throwing us back upon the tender mercies of Thomas Malthus.

How will these intertwined factors evolve in the years to come? There is always the unlikely possibility of changing human nature, but inventing, working, and thinking our way to a prosperous and sustainable future is a more achievable goal. Properly applied, science and engineering can offer solutions to humanity's ever-increasing need for energy, which might otherwise be our Achilles heel. The next four chapters give a planetary-scale view of the energy and climate problems we face over a time frame long in human terms, but short in the lifetime of the world on which we live.

Recommended Reading

Diamond, Jared. *Collapse: How Societies Choose to Fail or Succeed*. New York: Viking/Penguin, 2005.

Spengler, Oswald. *The Decline of the West*. Abridged edition. New York: Oxford University Press, 1926.

1
Ancient Days and Modern Times

> But I [Zeus] will give men as the price for fire an evil thing in
> which they may all be glad of heart while they embrace their
> own destruction.

Hesiod, *Works and Days* c. 800 B.C.

In the mythology of ancient Greece, fire was enshrouded in the legend of Prometheus, who stole its secret from Zeus and gave it to mankind. Because fire was not intended for us, humanity was punished for accepting it by the advent of Pandora, who opened the great box of woe, releasing disease, toil, and sorrow upon the world, saving only the gift of hope. This legend strikes a chord even now, as our use of energy expands seemingly without limit even as consequences begin to appear. Prometheus suffered greatly for giving fire to humanity, and we may yet suffer greatly from our overwhelming dependence upon it. Only time will tell if the gift of Prometheus turns out to be the blessing in the future that it has been in the past.

There is no limit to the amount of energy humanity may want. But there are certainly limits to the amount available from fossil fuels. Through the rise and fall of successive civilizations and empires, the energy needs of the ancient world were supplied by renewable sources, including food, firewood, wind, flowing water, draft animals, and slaves, with only tiny amounts from coal and oil. The 18th century witnessed the invention of steam engines that could turn heat into useful work. Because of this advance in technology, energy usage for the first time began to grow faster than population. The heat from burning coal powered the industrial revolution. Soon afterward it was discovered that devices such as steam engines are limited in the fraction of heat they can convert into work. By immutable laws of thermodynamics, steam turbines like those used in power plants typically waste more than half the energy they consume. Although work can be converted completely into heat, the reverse is not true. Heat cannot be converted completely into work, no matter how many engineering advances are made.

Now, at the beginning of the third millennium, the world's average annual energy consumption per person is about 100 times higher than it was 2000 years ago, when there were only perhaps 200 million people in the world. Presently, there are more than 6 billion souls on the face of the Earth. Over the last two

Energy Demand and Climate Change: Issues and Resolutions. Franklin Hadley Cocks
© 2009 WILEY-VCH Verlag GmbH & Co. KGaA, Weinheim
ISBN: 978-3-527-32446-0

millennia, world energy usage has risen by a factor of more than 3000 and has been increasing at a rate of around 2% per year. This 2% increase per year is all by itself 60 times more energy than the total annual energy consumption of the ancient world. The amount of energy used per person varies enormously from country to country. If all countries consumed as much energy per person as the richest do, world energy usage would be an order of magnitude higher than it is. Such an increase in energy may be unachievable with fossil fuels alone and, in any case, is not sustainable.

Almost everyone wants a higher standard of living. Energy and living standards go hand in hand. How can continuously increasing demands for energy be satisfied? Can energy use double in the next 40 years and keep increasing with no end in sight? Right now human energy needs are met primarily through the gift of Prometheus, from fires fed by fossil fuels. But all fossil fuels contain carbon, and burning it produces carbon dioxide. As the amount of carbon dioxide in the air increases, so does the Earth's average temperature, because carbon dioxide acts to slow down Earth's heat loss. Changes in global weather accompanying increasing planetary temperature will become greater as energy production from fossil fuels goes up.

Beyond fossil fuels, the energy contained in certain atomic nuclei can also be set loose, as was proven dramatically by the detonation of the first atomic bomb at Alamogordo, New Mexico, on 16 July 1945. Of more peaceful potential was the earlier proof that energy locked inside uranium atoms could be released controllably. It did not take long to demonstrate that the controlled release of nuclear energy can be used to generate electricity. Heat from the reactor built in 1943 in Oak Ridge, Tennessee, to demonstrate the production of plutonium from uranium, was used to generate electric power before the first nuclear weapon was detonated. (This reactor has long been decommissioned but still exists as a museum.) Although the amount of electric power produced was only symbolic, this test did prove the principle of the concept. Before the hazards and costs associated with nuclear reactors were fully realized, extravagant claims were made about the ability of this new energy source to supply electricity cheaply, notably the claim in the 1950s that electricity from nuclear reactors would become too inexpensive to meter. New technologies inevitably generate new problems as well as new possibilities, and the balance, if there is one, is always between cost and benefit.

The production of electrical energy using the heat from nuclear fission has increased far less rapidly than first expected. The combination of complex engineering, serious concerns about radioactive waste, and the role of reactors in making nuclear weapons have all inhibited the growth of nuclear-powered electricity generation. Where and how to store radioactive waste for millennia is not a simple problem. The politics of nuclear weapons are not simple either.

The impact of worldwide climate change needs scarcely to be emphasized. To understand why such climate changes are underway, it is important to know the scientific facts that determine our planet's temperature. While the details of this issue are extremely complicated, the fundamental principles are straightforward. The next chapter gives an overview of why ice ages and global warming cycles come and go. What happens if glaciers start growing again, the next ice age begins, and we're out of fuel?

Recommended Reading

Eberhart, Mark E. *Feeding the Fire: The Long History and Uncertain Future of Mankind's Energy Addiction.* New York: Harmony House, 2007.

Thirring, Hans. *Energy for Man: Windmills to Nuclear Power.* Bloomington, IN: Indiana University Press, 1958. (This book was one of the earliest to evaluate quantitatively humanity's energy situation, and much of its information is still relevant.)

2
Ice Ages – Past and Future

The glaciers creep like snakes that watch their prey,
from their far fountains, slow rolling on.

Percy Bysshe Shelley, *Mont Blanc*, 1816

In my opinion, the only way to account for all these facts and
relate them to known geological phenomena is to assume that at
the end of the geological epoch which preceded the uplifting of
the Alps, the earth was covered by a huge ice sheet …

Louis Agassiz, *Studies on Glaciers*, 1837

Global warming is now the question for our age, but this wasn't always so. At one time, the preeminent issue in Earth science was global cooling. The existence of glaciers has been known ever since humanity migrated toward the poles, but in 1837 the Swiss-American zoologist, geologist, and eventual Harvard professor Louis Agassiz (1807–1873) proposed the idea that glaciers could sometimes start growing and expand across large portions of the Earth's surface. What causes the Earth to cool and glaciers to advance across the land and then retreat again towards the poles? The answer lies partly in the Earth's periodic orbital changes, whose effects on climate were not understood until the 20th century. Even the very existence of times when the Earth was inundated with ice was not known until the 19th century.

The Discovery of Ice Ages

Louis Agassiz became convinced that the Earth had experienced periods in the past when glaciers were extremely large and widespread, covering most of Europe. Others had earlier entertained similar thoughts, but as president of the Swiss Society of Natural Sciences, Agassiz was in the right position to advance the theory he began to believe after observing deep scratches in the rocks near Neuchâtel in Switzerland. He concluded that these deep scratches could only have been made by the forced motion of boulders caught up in the moving ice of a glacier, which must have previously covered the area. This idea, that there have

Energy Demand and Climate Change: Issues and Resolutions. Franklin Hadley Cocks
© 2009 WILEY-VCH Verlag GmbH & Co. KGaA, Weinheim
ISBN: 978-3-527-32446-0

been times in the past when glaciers were far larger than they are now, was presented to the Society of Natural Sciences and immediately led to heated debate. Had he not made excessive claims for the maximum extent of these earlier ice sheets (that they extended to the Mediterranean and even further south, for example), the concept of ice ages might have been adopted sooner. But in the end, the idea that glaciers can advance and then retreat was accepted. Agassiz was one of those fortunate scientists who lived to see their ideas vindicated and accepted. He lies buried in Mt. Auburn Cemetery in Cambridge, Massachusetts. The tombstone he shares with his wife, Elizabeth, is a large boulder that had been transported by the Swiss glacier Aar to the area where he first observed the deep rock scratches that led him to conclude that glaciers wax and wane. The question then was posed: Why does the Earth sometimes heat up and then cool down again?

The Heat Balance of the Earth

Our planet has basked in the light of its yellow sun for more than five billion years. All this long while, the Earth's temperature has been determined by the balance between the energy it receives and the energy it loses. The Earth's energy input is delivered almost entirely by direct sunlight, which is 450,000 times more than we receive from moonlight and about 5000 times greater than the heat diffusing up from the Earth's core. Its energy loss is entirely in the form of the infrared radiation it emits. The fraction of incident sunlight that is reflected is called the *albedo*, from the Latin word *albus*, meaning white, and it varies as atmospheric conditions change. When white clouds are especially numerous across the face of the Earth, reflected sunlight increases sharply as the albedo increases. Sunlight reflected from the Earth can sometimes be strong enough to make visible the outline of the darkened portion of the moon within the arms of the bright lunar crescent. The renowned Italian renaissance painter, architect, and engineer Leonardo da Vinci (1452–1519) may have been the first to suggest that earthlight gives rise to this lunar phenomenon. Averaged over time, the Earth reflects about one-third of incident sunlight and absorbs about two-thirds.

Earth's heat loss is entirely in the form of infrared radiation, like the heat you can feel on your hand but can't see with your eyes when the electric coils on a stove are still warm but no longer glow red. Clouds affect the amount of infrared heat radiation escaping from the Earth as well as the amount of sunlight reaching its surface. Our current atmosphere prevents roughly one-third of the Earth's infrared radiation from escaping. As long as the average absorptivity of the atmosphere for infrared radiation and the average albedo of the Earth stay constant, why should the average temperature vary?

The world is just a ball of matter with an onionskin of atmosphere, orbiting its yellow sun in the vacuum of space. The heat balance of the Earth is simple enough in principle, though complicated in its details. When the Earth's temperature is stable, the heat coming in exactly equals the heat going out. If the heat received from the sun is more than the infrared heat the Earth radiates into space, the global temperature will rise. If the heat radiated away exceeds the heat received, the temperature will fall. But arguments abound concerning precisely how and why this planetary balance can shift independently of the coming and going of clouds.

The balance between the energy reaching or leaving the Earth must be upset somehow if the average temperature varies. What can cause this heat balance to change? There are many possible explanations, including the sun itself, the orbit of the Earth, and the angle at which our planet spins about on its axis. Clouds, dust, carbon dioxide (CO_2), and other gases in the atmosphere, including water vapor (H_2O) and methane (CH_4), are also important. There is an enormous amount of water vapor in our atmosphere but only very small amounts of methane or carbon dioxide. Even with the dramatic increase in CO_2 that has occurred in the last 100 years, there is still over 20 times more argon, the inert gas used in fluorescent lights, than carbon dioxide in the air. The very low level of CO_2 compared to that of oxygen and nitrogen makes it possible for humanity to significantly increase its atmospheric concentration. As it turns out, both CO_2 and CH_4 can affect global temperature, even at their very low levels. Some of the other factors that determine planetary temperature occur in predictable cycles, such as those caused by periodic changes in the Earth's orbit. Still other things that might affect the Earth's temperature are not very predictable. Let's begin with the sun itself.

The Sun and Its Spots

You might think that the energy output of the sun is constant, but this is not so. Over billions of years the nuclear fires that power the sun are slowly changing, and the sun is gradually growing hotter. But this effect is extremely slow indeed. Other things being equal, the increasing energy output of the sun will raise the Earth's temperature only half a degree centigrade in about 50 million years. The appearance and disappearance of dark spots on the face of the sun is a much shorter-term effect. On the sun's surface, dark spots come and go in cycles lasting approximately 11 years, as shown in Figure 2.1.

These dark spots look black only in contrast to the brilliance that surrounds them. They can be of enormous size, much larger than Earth itself. Figure 2.2 is an image of the sun taken in August 1999 by the Solar Heliospheric Observatory (SOHO) satellite. In this case the sunspots appear along the mid-latitudes of the sun's northern and southern hemispheres. The sun is not solid and rotates most rapidly at its equator and more slowly towards its poles.

Surprisingly, the total light from the sun is higher when several dark sunspots are present on its surface than when there are not very many. Why is this so? It turns out that the surface of the sun, especially around sunspots, is hotter in local regions (termed *faculae*) on the sun's surface. Inside sunspots, the solar surface is cooler than usual–which is why they appear dark–but the hotter faculae in the margins around sunspots more than make up for their cooler interiors. Precisely

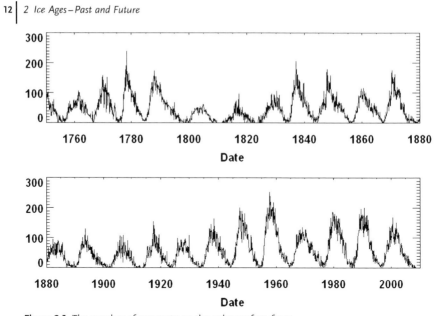

Figure 2.1 The number of sunspots on the solar surface from before 1760 to after the year 2000, revealing clearly the cyclical nature of these spots.

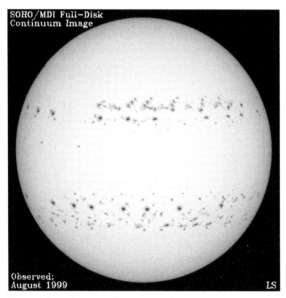

Figure 2.2 Appearance of the sun in August 1999, as seen in a SOHO continuum image.

what sunspots are is not completely clear, but one conceptual view is to think of them as the eyes of magnetic hurricanes, with the dark spots formed by concentrated magnetic fields that block heat flow up from the sun's core. The hot faculae writhing in the weaker magnetic fields around sunspots give off heat so strongly that the total energy radiated, including the spot itself plus the area in its margin, is actually greater than that from an unblemished area of the same size. This is why the sun's overall energy output is higher when sunspots are present. When the face of the sun is relatively free of spots, the energy of sunlight is about 0.05% to 0.07% less than if there are lots of sunspots. Superimposed on this sunspot effect is the day-to-day variation, up to 0.3%, caused by sporadic changes in the rate at which the sun's internal heat percolates outward from its interior, much as some coffee makers produce chaotic bubbling noises that vary in intensity.

This sunspot cycle has been observed for centuries, except for a period of 70 years (1645–1715), when spots nearly disappeared. Nobody knows why there were so few sunspots for such an unusually long period of time or whether this was a random or a repeating event. This missing-sunspot period is called the Maunder minimum, after the English astronomer Edward Maunder (1851–1928), who was the first to recognize this effect in the historical record. Maunder also observed that high sunspot activity corresponded to high auroral activity, and solar flares and sunspots tend to occur together. Auroras, the glowing curtains of light that appear around the North and South Poles, are a result of energetic particles cast out from the sun by solar flares getting caught in the web of the Earth's magnetic field and spiraling around in it until hitting the atmosphere high above the Earth's magnetic poles. The impact of these particles causes the upper air to glow, while also generating radio static and other electrical phenomena.

The effect of sunspots is too small to change the Earth's energy balance enough to cause ice ages, but some have argued that the prolonged decrease in the solar energy reaching the Earth during the Maunder sunspot minimum worsened the so-called Little Ice Age. This global decrease in temperature began in the 14th century and only faded out at the end of the 19th century. Although glaciers did increase in size in many parts of the world during this period, new, continent-covering glaciers did not appear. Interestingly, the Little Ice Age started just after decades of warmer weather, known as the Medieval Warming Period, when food production and population increased, especially in the Northern Hemisphere. Greater population accentuated the difficulties brought on by the onset of colder weather, which depressed food production. The Flemish artist Pieter Bruegel (1525–1569) captured well the bleak flavor of winters during the Little Ice Age in his painting *Hunters in the Snow*, reproduced in black and white in Figure 2.3.

There is no completely satisfactory explanation for either the initial warming or the subsequent Little Ice Age in which Bruegel lived. However, there are several major climate-influencing effects that, combined, are large enough to explain why ice ages occur, beginning with changes in the Earth's orbit around the sun.

Figure 2.3 *Hunters in the Snow*, a winter scene in Europe during the Little Ice Age as captured in a painting on a wooden panel in 1565 by the Flemish artist Pieter Bruegel the Elder (c. 1525–1569).

Earth's Orbit

Even small changes in the orbit of the Earth and the angle at which its poles are inclined with respect to the plane of this orbit cause significant variations in sunlight at the poles. Orbit changes and the tilt of its poles, when added together, have played a major role in the sequence of true ice ages over the last million years. In major cold periods, glaciers covered large parts of the Northern Hemisphere. Over a much longer time frame (hundreds of millions of years), ice ages may also be related to the drifting of continents. Gigantic ice sheets form more easily over land than over water. In the current geological epoch, there is a larger fraction of the Earth's land surface in the Northern Hemisphere than in the Southern Hemisphere, but this may not have always been the case. Because most of the landmass of the Earth is now north of the equator, extending beyond the Arctic Circle, the formation of new, massive glaciers during the ice ages of the last few million years occurred mainly in the Northern Hemisphere. Continents move very slowly, and this movement played no part in the ice ages that have come and gone over the last thousand millennia. Just why did massive ice sheets periodically sweep across much of the Northern Hemisphere? It turns out that the ice ages occurring during the last one million years were caused by changes in the way the Earth tilts, wobbles, and orbits the sun. To understand this effect, let's look at some basic facts about the solar system we inhabit.

The Earth orbits the sun in about 365 days, while Jupiter, the largest planet, takes approximately 4300 days. Saturn, which is further away than Jupiter, orbits in about 10,800 days. Because all planets orbit the sun in the same direction, the gravitational tugs that Jupiter and Saturn exert on the Earth repeat in a little more than a year. Every few decades the gravity tugs from both Jupiter and Saturn nearly coincide. Mars and Venus approach closer to us than Jupiter does, but together they have only about 0.3% of its mass, so their gravitational effect is very small. The force of gravity increases in proportion to mass and decreases as the square of the separation distance. The gravitational effect of Jupiter on Earth is by far the largest, followed by that of Saturn. The cyclical gravitational effects of Jupiter and Saturn slowly cause the orbit of the Earth to distend periodically to a maximum distance that is farther from the sun than is presently the case. Kepler and Newton first provided the basic facts needed to explain this orbital motion.

The Discovery of Elliptical Orbits

The German astronomer Johannes Kepler (1571–1630) discovered that the planets move around the sun in elliptical orbits, not circles, and certainly not in the so-called epicycles, that is, multiple circles, proposed in the ancient system of the Alexandrian astronomer and mathematician Ptolemy (circa 150 A.D.). The whole epicycle business was the ancient way of explaining the retrograde motion of the planets, those brief times when they appear to move backwards in the sky. Kepler's efforts to explain planetary motion were handicapped by the mediocre data he had at hand, but he knew that the best data available were those of the Danish nobleman Tycho Brahe (1546–1601). Brahe had assembled an enormous collection of planetary position measurements. His island observatory in Denmark, the Uranienborg ("castle of the sky"), was the best in Europe, and he was the last great astronomer to work by naked-eye measurements alone.

After Brahe was forced by the new King of Denmark to leave the Uranienborg and he moved to Prague, Kepler was able to get a job as his assistant. Even then Brahe kept his measurements to himself. Only after Brahe's death did Kepler gain access to his data. With Brahe's extensive collection of observations extending over many years, Kepler was able to show that the motion of the planets, including retrograde motion, could be explained if they moved around the sun in ellipses, which are distorted circles elongated in one direction. Kepler's first law of planetary motion states this fact, and his other two laws are consequences of elliptical planetary orbits. Kepler's discovery of the elliptical orbit of planets was the first step on the long road leading to the explanation of ice ages.

Over time, the repeating cycle of gravitational tugs from Jupiter and Saturn make the Earth's orbit slightly more elliptical and then less elliptical again This whole process of orbit stretching and contracting repeats about every 100,000 years, with possibly a smaller periodic distortion every 400,000 years. Figure 2.4 illustrates schematically the elliptical nature of Earth's orbit.

With the elliptical orbit that the Earth currently has, its closet approach to the sun (perihelion) occurs in the winter of the Northern Hemisphere, when it is about 3.4% closer than it is in the summer. At this closest approach, the Earth receives

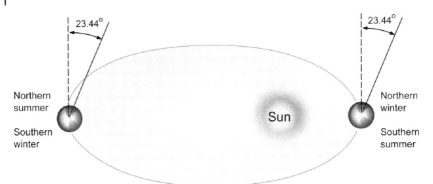

Figure 2.4 The elliptical nature of the eccentric orbit of the Earth around the sun, in exaggerated fashion. The eccentricity of the orbit as shown is about 60 times that of Earth's actual current orbit. For our purposes it is sufficient to note that eccentricity is a quantitative measure of the extent to which a circle is distorted. Currently, the eccentricity of Earth's orbit is 0.0167. If Earth's orbit were circular, its eccentricity would be zero, and the sun would be at the center. This 0.0167 deviation from circularity is so small that for a drawing on the scale of this book, it is visually indistinguishable from a circle. For the solar system, an eccentricity of 0.0167 means the sun is not at the center, as the orbit is elliptical, not circular. Because of this elliptical orbit, the Earth is about three million miles (approximately 5 million kilometers) closer to the sun at its closest approach (presently 3 January) than it is at its farthest point (presently 4 July). Over about 100,000 years the eccentricity of the Earth's orbit varies from about 0.0034 to almost 0.058 because of the gravitational influence of Jupiter and Saturn. As a result, the differences between Earth's maximum and minimum distances from the sun vary by more than a factor of 10. This large variation is a major factor in the coming and going of ice ages because of its effect on the amount of energy received from summer sunlight in the polar regions.

about 6.8% more sunlight than when it is furthest away from the sun (aphelion). In millennia to come, when the Earth's orbit is at its maximum elliptical extension, the amount of sunlight received at perihelion will be more than 20% greater than at aphelion. Earth's elliptical orbit changes have major implications for the heat balance of the Earth, but they are not great enough to cause ice ages by themselves. Something more is needed, and, indeed, there are other astronomical influences that interact with the Earth's periodic orbital distortions. One of these additional effects is precession.

Precession

At the same time that orbital stretching changes are going on, the Earth's rotation axis also swings slowly around in a motion called *precession*. Like the axis of a gigantic spinning top, the Earth's rotation axis very slowly sweeps out in an enormous cone shape in space. The current tilt of the Earth's rotation axis with respect to the plane of its orbit is shown in Figure 2.5.

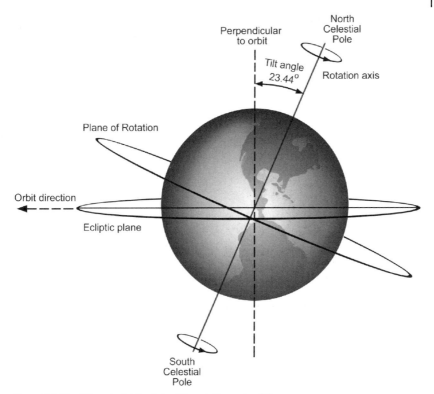

Figure 2.5 The tilt angle (obliquity) of the rotation axis of the Earth with respect to the plane of its orbit (the ecliptic). The angle of this tilt is currently about 23.44 degrees. The rotation axis of the Earth swings around in space over a period of 25,725 years, a process called precession. Simultaneously, over a period of 41,000 years, this tilt angle cycles slowly between 21.1 and 24.5 degrees.

Because of precession, Earth's rotation axis points over time to different portions of the sky. The swinging-around action of the Earth's axis is similar to that of a spinning top, but the causes are not analogous. For a spinning top, it is the downward force of gravity that causes its spin-axis to swing around. The basic gravitational reasons for the Earth's precession were deduced by Isaac Newton, who at age 45 described the law of gravity in his immortal book, published in 1687, *The Principles of Natural Philosophy* (the Latin title is traditionally shortened to *Principia*). In this book, Newton demonstrated how Kepler's discovery of the elliptical nature of planetary orbits can be deduced from the law of gravitation. Newton was also able to show that the precession of the Earth's axis of rotation originates from the fact that the Earth is not perfectly spherical. (Newton was a better mathematician than metallurgist. If he had remained at Cambridge University and declined appointment as Master of the Mint, very likely he would have avoided impairing his remarkable mathematical brain in attempting to turn lead into silver by marinating it in mercury, as a means of solving England's money problems.)

Added to the changing elliptical motion, the slow swinging around of the north poles of the Earth's rotation axis also affects global weather and contributes to the ice age cycle. Our North Polar axis currently points toward the star Polaris (the North Star). In 12,000 B.C. the North Pole pointed toward the star Vega, and in 14,000 A.D. it will point to Vega again. One complete precession of the rotation axis of the Earth takes 25,725 years, about one-quarter of the 100,000-year orbital ellipse elongation and contraction cycle. These two effects, the stretching and shrinking of the elliptical shape of the Earth's orbit and precession, which slowly changes the direction of the Earth's tilt angle, are the next steps in explaining why ice ages occur. Changing seasons are the result of the tilt of the Earth's axis, which remains virtually fixed as the Earth orbits the sun from one year to the next, as seen in Figure 2.4.

Because of this tilt, the Northern Hemisphere receives more hours of sunlight in the summer than it does in winter. As the Earth moves around its orbit, there are times in the spring and the fall when the length of days and nights is equal. These so-called equinoxes now occur around 21 March and 22 September. Because of precession, the dates at which equinoxes occur change over time, a process termed the precession of the equinoxes. Surprisingly, one precession of the equinoxes takes about 23,000 years, not the 25,725 years needed for Earth's spin axis to rotate completely around. About 5500 years ago, for example, the seasons were shifted one-quarter of a year from where they are now, and the equinoxes occurred around 21 December and 21 June. What causes this 23,000-year cycle? It turns out that the long direction of the slowly changing ellipse of the Earth's orbit (See Figure 2.4) itself rotates slowly around the sun. The rotation of the major axis of Earth's elliptical orbit is in a direction opposite to the 25,725-year swinging around in space of the rotation axis of the Earth (precession). The combination of these two rotation effects reduces the time required for the precession of the equinoxes from 25,725 to about 23,000 years and also gives rise to a smaller cyclic effect that occurs every 19,000 years. But even the Earth's changing elliptical orbit and the precession of its rotation axis added together are still not enough to explain completely the occurrence of ice ages. There is yet another important factor, and that factor is nutation.

Nutation (Wobble)

As the Earth spins on its rotation axis, this axis also wobbles slowly because of the gravitational force of the moon. So, while the Earth is spinning like a top and its orbit is stretching and contracting as its axis of rotation is slowly precessing around in space, the angle at which the Earth's rotation axis is tilted is also changing. Clearly, explaining the motion of the Earth is not simple. The wobbling of the Earth's tilt axis (the technical term is *nutation*) occurs because the Earth is actually fatter around its equator, like a man with a potbelly, and this extra mass is acted upon by the gravity of both the moon and the sun. The gravitational force acting on this girdle of matter around the Earth's middle causes the rotation axis to swing slowly around and also to wobble, just as the downward force of gravity causes the rotation axis of a spinning top to simultaneously swing around (precess) and wobble (nutate).

The nutation of the Earth's axis of rotation, that is, the upward and downward wobble of its rotation axis, is not large but would be much smaller still if our moon were not as big as it is. The mass of our moon is about 1.2% that of the Earth and is very large as moon–planet combinations go. The mass of both of Mars' moons added together is less than one ten-millionth the mass of their host planet, for example. The wobble of Earth's axis caused by our large moon changes the angle (the *obliquity*) that the axis of rotation of the spinning Earth makes with the plane (the *ecliptic*) in which the Earth revolves around the sun (see Figure 2.5). Right now, the axis of the Earth is inclined about 23.44 degrees to the ecliptic. As the Earth goes around its orbit, the daylight side of the Northern Hemisphere is tilted 23.44 degrees away from the sun in winter, while in summer it is tilted the same amount toward the sun.

Over a period of about 41,000 years, this wobble causes the tilt angle (obliquity) to vary slowly from approximately 21.1 to about 24.5 degrees and then start back again. As the tilt angle of the Earth's rotation axis becomes smaller, the noontime summer sun at the poles moves closer to the horizon and shadows grow longer, which tends to make summers at the poles somewhat colder. At the Arctic and Antarctic Circles, the sun does not set for at least one day during the summer or rise for at least one day during the winter. As the tilt angle becomes larger, summers at the North Pole and South Pole grow warmer as the noontime summer sun rises higher up into the sky and noontime shadows grow shorter, making polar summers warmer. How do all these effects added together cause ice ages? The answer is in the effect they have on the degree to which ice is preserved year round in the polar regions, which changes a lot as the Earth's tilt angle, orbit, and wobble change a little.

It should be no surprise that changes in the tilt of the Earth are most dramatic around the poles. After all, the polar regions also show the greatest seasonal climate changes. Near the equator, differences in temperature between winter and summer are small. Near the poles, temperature changes between winter and summer are large. The influence of the changing elliptical nature of the Earth's orbit and the precession and wobble of its axis of rotation, when combined, has a large enough effect on summer sunlight in the polar regions to explain ice ages. But it has taken a long time to figure out how all these effects work to change global climate the way they do to produce or end ice ages, or even that ice ages occur at all.

Glaciers can grow and then shrink because of tilt axis wobble, precession, and changes in the elliptical nature of the Earth's orbit. It is all of these effects in combination that produce or end ice ages. Not all ice ages or all warming cycles are the same, of course, as wobbling, tilting, and orbital stretching repeat over different time periods. Right now, for example, the combination of wobble, elliptical orbit stretching, and precession combine to give the polar regions more solar energy than they will have in times to come. But within 2000 years, polar glaciers could begin to grow again. In addition to all these predictable orbit, precession, and wobble effects, there is another major but unpredictable natural phenomenon that also affects climate. This factor is volcanic dust.

Volcanic Dust

Major volcanic eruptions happen every now and then, and the resulting dust these eruptions add to the atmosphere can affect temperatures by changing planetary albedo, the fraction of sunlight the Earth reflects back into space. In 1883, for example, the dust level in the upper atmosphere was suddenly raised when the gigantic eruption of Krakatoa, a volcano on a small islet in Indonesia between Java and Sumatra, propelled volcanic ash high into the stratosphere. The resulting increase in planetary albedo led to a noticeable, but temporary, worldwide drop in temperature Even earlier, in 1815, the far larger explosion of the volcano Tambora on the Indonesian island of Sumbawa threw 30 cubic miles (150 cubic kilometers) of dust high into the atmosphere. The resulting drop in temperature caused the so-called "year without a summer," because this upper atmosphere dust reflected so much sunlight back into space. This eruption was 100 times the size of Vesuvius or Mt. Saint Helens. But even the Tambora eruption pales in comparison to the eruption 75,000 years ago of Mount Toba on the Indonesian island of Sumatra. This cataclysmic event hurled nearly 700 cubic miles (2800 cubic kilometers) of dust into the atmosphere. Volcanic dust, however, even vast amounts of it, normally floats around in the upper atmosphere for only a few years, unlike Earth's orbital effects, which endure for millennia. And volcanic eruptions are not cyclical, as ice ages are.

The Cyclical Nature of Ice Ages

The complexity of the orbital effects of the Earth is considerable, but nonetheless they provide the first good explanation of why ice ages come and go. The possibility that changes in the tilt axis of the Earth could cause ice ages was initially proposed in 1842 by the French mathematician Joseph Adhémar (1797–1862) in his book *Revolutions of the Sea*. In his time, the fact that the tilt of the Earth changes slowly had already been known for nearly 2000 years. The Greek mathematician Hipparchus (circa 120 B.C.) deduced this change in tilt by comparing his own astronomical measurements with earlier ones extending back to those made in Babylonian times. Adhémar, however, drew wildly extravagant conclusions, suggesting, for example, that the gravitational attraction of the Antarctic ice sheet during an ice age would drain large amounts of water from the oceans of the Northern Hemisphere. He ludicrously concluded that this Antarctic ice sheet would grow to such a size that its eventual melting would form it into a gigantic mushroom shape, whose collapse would cause massive flooding on a planetary scale. Because of these ridiculous scenarios, his whole concept was ridiculed and dismissed. But a small grain of truth remained in his concept, and a determined man named James Croll built upon this tiny grain.

The Croll–Milanković Theory of Ice Ages

James Croll (1821–1890) was the impoverished son of a Scottish stonemason and earned his early living as a mill worker, innkeeper, and janitor. He had no formal higher education and was self-taught in physics and mathematics. In his geophysical studies, he was able to quantify some of the effects that changes in the Earth's orbit, tilt, and wobble have on the sunlight received in polar regions. Croll was not entirely correct in his theory of ice ages, and one error that he made was his assumption that massive ice sheets would alternate between the Northern and Southern Hemispheres. The Serbian mathematician Milutin Milanković (1879–1958) subsequently was able to work out in detail the mathematics governing the effect of orbit, tilt, and wobble on the sunlight received by the Earth.

Milanković made quantitative calculations of the amount of solar energy received per unit area of the Earth's surface (called the *insolation*) as a function of latitude around the North and South Poles over the last 650,000 years. In the days before computers, these calculations of insolation values for different latitudes had to be done by hand, and Milanković worked more than 30 years to complete them. During this time he also became entangled in the Balkan Wars and for a while was a prisoner in Turkey during World War I. His calculations showed that the changes in the amount of solar energy received at latitudes above 65 degrees, caused by orbit, tilt, and wobble effects, were large enough to induce the ice ages as revealed by the geological record. Unlike Croll, Milanković believed, correctly, that lower summer temperatures in the polar regions were more important than lower winter temperatures, because cold summers preserve the most ice and snow from one year to the next.

At first, Milanković's explanation was hailed as a breakthrough, but then disputes arose among geologists concerning the exact cyclical nature of ice ages, and these arguments cast doubt on the whole orbital business. Advances in deep-sea drilling and seabed coring in the second half of the 20th century produced ocean sedimentation data that gave evidence for the sequence of dates for ice ages that coincided with those of the Croll–Milanković theory, at least over the last one million years. Ice and seabed cores, added to the data from ground cores, also give evidence for climate perturbations in the Little Ice Age as well as true ice ages. Some of these climatic disturbances were quite large, perhaps the most famous being the so-called younger Dryas cooling period that occurred about 12,000 years ago, 1000 years after the glaciers had first begun retreating. During the younger Dryas time, the climate turned several degrees colder. For more than 1000 years the forests that had advanced into northern Europe were replaced once again by tundra as the glaciers began advancing. In this colder climate, the alpine wild flower *Dryas octopetala* reappeared, leaving traces in core samples that are younger than the ones formed in the days of ice ages.

Rapid climate changes like the younger Dryas event are not on the same scale as actual ice ages. Many theories have been proposed to explain these smaller events, and arguments still abound, but no theory is as accepted as the explanation of the cyclical nature of the ice ages that have occurred during the last one million years. Current thinking is that the younger Dryas cooling may have started when a giant lake that formed in North America from glacier melt water at the end of the last ice age suddenly drained, adding enormous quantities of fresh water to the North Atlantic, thereby shutting down the Gulf Stream. Others have suggested that the younger Dryas cooling may have been caused by a meteor impact, like the one that eliminated the dinosaurs. In the end, nobody knows for sure what happened to cause this temporary cooling, but the cause of genuine ice ages is now understood.

Final proof of the relation of ice ages to Earth's orbit, tilt, and wobble motions came with the discovery that the Earth's magnetic field can shift dramatically from time to time. In 1906, four years before his death at 39, the French natural scientist Bernard Brunhes published his discovery that the weak magnetic field found in bricks correlates with the direction of the Earth's magnetic field in which these bricks were cooled after firing. He also discovered that the magnetic field found in lava after it has solidified behaves in the same way. Building on Brunhes' work, the Japanese paleomagnetist Motonori Matuyama (1884–1958) established, through numerous studies of the magnetic polarity of rock deposits, that the Earth's magnetic field must somehow periodically reverse direction. It was this observation that provided a method for correlating the dates of ice ages as recorded in the geology of both the Northern and Southern Hemispheres. The periodic reversal of the Earth's magnetic field preserved in volcanic rock showed that the ice ages on the northern and southern continents were synchronized. This correlation gave the final proof that these ice ages were related to periodic changes in the Earth's orbit, tilt, and wobble.

It took more than 50 years of geological observations to prove that the reversal of the magnetic field was global, not local. Because of the global nature of the Earth's magnetic field, it became possible to correlate the occurrence of glaciations on different continents and to show that these glaciations coincide neatly, on all continents and in the sediments on the ocean floor and in the ice layers in Antarctica and Greenland, with the predictions of the Croll–Milanković theory. More than 100 years after Louis Agassiz hypothesized that ice ages occur, the reasons why they happen have been proven to relate to changes in the Earth's orbit, tilt, and wobble. Sadly, both Croll and Milanković died without having the satisfaction that Agassiz had, of seeing their theory proven correct and fully accepted. With global warming now underway, what do the Croll–Milanković theory and Earth's finite fossil fuel supply portend with respect to the onset of the next ice age? The next chapter addresses this question.

Recommended Reading

Alley, Richard B. *The Two Mile Time Machine*. Princeton, NJ: Princeton University Press, 2000.

Imbrie, John and Katherine Palmer Imbrie. *Ice Ages: Solving the Mystery*. Short Hills, NJ: Enslow, 1979.

Macdougall, Doug. *Frozen Earth: The Once and Future Story of Ice Ages*. Berkeley, CA: University of California Press, 2004.

Milanković, Milutin. *Canon of Insolation and the Ice-Age Problem*, Spec. Pub. 132 (Section of Mathematical and Natural Sciences, 33). Belgrade: Royal Serbian Academy, 1941. Trans: Jerusalem: Israel Program for Scientific Translations, 1969. National Technical Information Center, document TT-6751410/1 and 2.

Newton, Isaac. *The Principia*. (Mathematical Principles of Natural Philosophy). Trans. Andrew Motte. Amherst, NY: Prometheus Books, 1995. See also: http://en.wikipedia.org/wiki/Philosophiae_Naturalis_Principia_Mathematica

3
Global Warming Versus Returning Glaciers

If now the quantity of carbonic acid in the air is increased, the
temperature of the earth's surface increases. This also causes
some increase in the amount of aqueous vapor in the air,
resulting in a slight intensification of the effect.

Svante Arrhenius, *Nature's Heat Usage*, 1896

Despite global warming, the Earth is still entangled in the ice age cycle and is slowly approaching the time when glaciers are scheduled to return in force. Thanks to Earth's orbit, tilt, and wobble changes, we might be within about 2000 years of the end of the current era of mild climate. Does this mean that global warming may save us from returning glaciers? Don't bet on it. To begin with, global warming goes hand in hand with higher levels of greenhouse gases. Unusually high levels of these gases are slowly purged from the air by natural processes, such as the dissolution of CO_2 into the oceans or the formation of carbonate rocks, in times measured in centuries and millennia, not hundreds of thousands of years. Meanwhile, as greenhouse gases cause warming, it can be argued that they spur the growth of central Antarctic glaciers, while at the same time eliminating other, non-polar ice masses. The best current estimate is that over the next 100 years, the Earth's average temperature will increase by 3.2–7.2 °F (1.8–4 °C), and the oceans will rise by 7–23 inches (18–58 centimeters).

How could warmer weather speed up the growth of Antarctic glaciers? One mechanism is through increased snowfall. As the temperature rises, the rate of water evaporation speeds up dramatically, as anyone who has dried laundry outside in wintertime knows. Raising the temperature of water from 77 to 78.8 °F (from 25 to 26 °C) enhances evaporation by about 6%, for example. Increasing the surface ocean temperature over the next century might increase its evaporation rate more than 20%! Also, as the temperature increases, air has the ability to hold more water. Cold air can retain less water than warm air can, which is why cold fronts can cause rain. The water that evaporates from the oceans and from the land eventually precipitates out somewhere as rain, hail, sleet, snow, dew, or frost. But predicting exactly where this precipitation will happen is a major problem in meteorology.

Energy Demand and Climate Change: Issues and Resolutions. Franklin Hadley Cocks
© 2009 WILEY-VCH Verlag GmbH & Co. KGaA, Weinheim
ISBN: 978-3-527-32446-0

Nobody suggests that global warming will raise the temperature in Antarctica above freezing in winter. Increased atmospheric humidity due to enhanced evaporation from the oceans has already led to higher snowfall across central Antarctica. Normally, Antarctic air is extremely dry, drier even than desert air in Arabia. What little snow falls in Antarctica, however, is mostly retained from one season to the next. A little snow sublimes, turning directly from solid ice into vapor without melting, just as the ice cubes left in a freezer tray slowly decrease in size over a long time due to the same solid-to-vapor process, but this effect is small compared to increased snowfall. As ocean temperatures rise, more water evaporates, and in Antarctica snowfalls could reach levels higher than they are now. Growth of the central glaciers in Antarctica counteracts the increase in the volume of ocean water resulting from higher water temperatures and the melting of non-Antarctic ice. Currently, the total mass of Antarctic ice is actually growing larger due to increased snowfall, even as big icebergs keep breaking off, especially from the West Antarctic ice sheet, which extends below sea level and shows signs of breaking up. At the same time, Arctic ice is decreasing, and the snows of Mt. Kilimanjaro and most of the non-polar glaciers in the Alps, New Zealand, Canada, South America, and Alaska are visibly receding as year-round temperatures increase. The accumulation of more ice and snow in central Antarctica, at least for the next 100 years, is not expected to outdo the melting of Arctic ice and non-polar glaciers or the expansion of warming ocean water, so the ocean level will keep rising.

Most Arctic ice is floating on water. When floating ice melts, the ocean level does not increase, because floating ice displaces as much water as it produces by melting. The warming of the air and oceans appears to be melting floating Arctic ice during summer from both the top and bottom. The decrease in floating Arctic ice exposes increasing areas of the Arctic Ocean to sunlight. Water absorbs much more sunlight than snow and ice do, and the higher evaporation rate from ice-free seawater might cause greater snowfall across the northern polar regions, producing larger snowfields that reflect more sunlight. Could the snowfall become so great that ground snow cover could persist through the brief northern summer at lower latitudes? Nobody knows for sure.

Higher planetary temperatures and melting ice might alter major ocean currents, especially the Gulf Stream. The influx of large quantities of fresh water from melting glaciers in Greenland would lower the salinity of northern oceans, especially in the North Atlantic, which in turn could change the northward flow pattern of salty, tropical water and reroute its return path through the deep ocean back toward the equator. The disappearance of the Gulf Stream would make large parts of Europe much colder, even though the west-to-east wind pattern helps in keeping the weather in northern Europe milder than in similar latitudes in eastern North America. Whether global warming will finally end up increasing or decreasing glaciers in Antarctica, there is little consensus as to exactly what changes at the South Pole might mean for the global climate.

There is no denying the fact that the Earth is warming up. Since 1900, global temperature data show convincingly that the Earth's average temperature has risen at least 1.3 °F (0.7 °C) over the 20th century, and temperature increases over the 21st

century will likely be from 2.5 to about 5 times higher. Why is this change in global climate happening? Let's take a look first at the basic scientific question of the rate of energy leaving the Earth and how the composition of the atmosphere affects it. It turns out that the transparency of the atmosphere to infrared radiation is a critical factor.

Infrared Radiation and Absolute Temperature

The Austrian physicist Ludwig Boltzmann (1844–1906) was an outstanding but tragic figure in the science of thermodynamics. He was able to prove, by mathematical reasoning alone, that the rate at which energy radiates from the surface of anything increases as the fourth power of its absolute temperature. This result turns out to be of great importance in understanding global warming. Boltzmann combined statistical mechanics with probability theory to deduce the fundamental thermodynamic laws. Unfortunately, his hidebound thermodynamic contemporaries disparaged his pioneering work, so that, suffering from illness and depression, he took his own life. His tombstone in Vienna is engraved with the equation $S = k \log W$, which he deduced in 1896 for the relation between the logarithm of probability, W, and entropy, S, a fundamental factor relating to the unavailable energy in a thermodynamic system. In this equation, the proportionality factor, k, between these two is now known as the Boltzmann constant.

The absolute temperature scale that Boltzmann used in determining the temperature dependence of emitted radiation is based on the fact that the temperature of anything can never be less than −459.67 °F (−273.15 °C). This lowest possible temperature is defined as absolute zero, or the temperature at which molecular motion ceases. Above absolute zero, Boltzmann showed that everything, including people, the Earth, and all else, emits radiation in rapidly increasing amounts as the temperature rises.

We have already mentioned in Chapter 2 how the very strong dependence of radiated energy on absolute temperature affects the output of the sun during the sunspot cycle. For things that have a temperature of less than about 1000 °F (538 °C), the energy that is radiated away is almost entirely invisible infrared heat. Electric heating coils on a stove appear dark even though they can still burn you. The current average temperature of the Earth is about 59 °F (15 °C). At this temperature the infrared wavelength it emits at the greatest rate is about 10,000 nanometers (0.00001 meters), approximately 20 times the wavelength of visible light and far beyond the range of human vision. The wavelength the sun emits at its maximum rate is about 460 nanometers, near the maximum sensitivity portion of the human vision range (green light), which is not surprising because our eyes have evolved over millions of years to utilize sunlight efficiently.

The vast bulk of the sun's energy is delivered at wavelengths within the range of 100 to 1000 nanometers. This range includes both ultraviolet light (about 400 nanometers or less) and infrared light (about 700 nanometers or more). The human eye can only see light in between these two limits, 400 to 700 nanometers, so a large part of the light from the sun is invisible to us. How was this invisible light discovered? Two developments made this discovery possible. Newton observed that a prism could separate sunlight into a rainbow of colors and that a second prism could restore these colors again to normal white light, as described in his book on optics (published in 1704 when he was 62). Galileo is credited with inventing the thermometer in 1592 when he was 27 years old (it was not until 17 years later that he turned a telescope toward the heavens). Combining Newton's prism with Galileo's thermometer, others attempted to measure the temperatures generated by the different colors of light. Surprisingly, it was discovered that the temperature in the rainbow spectrum produced by a prism kept going up even beyond the last visible red light of the solar spectrum. There is a lot of energy in the invisible, infrared portion of the solar spectrum, and this invisible part of sunlight helps to warm the Earth (and old men sleeping in the sun), even though it cannot be seen. In a similar fashion, the invisible long-wavelength (around and above 10,000 nanometers) infrared radiation given off by the Earth carries energy away from the Earth. To do this, however, it must first pass through the atmosphere, which itself radiates infrared heat. This is where carbon dioxide (CO_2), water vapor (H_2O), and methane (CH_4) affect the Earth's temperature. All of these gases, as well as some others, such as the chlorofluorocarbons used as refrigerants and nitrous oxide from automobile exhaust, strongly absorb portions of the infrared radiation emitted from the Earth and prevent its escaping into space. As we shall see, the fact that these gases absorb somewhat different portions of infrared light turns out to be critically important.

Greenhouse Gases and Global Warming: Fourier, Tyndall, and Arrhenius

Jean-Baptiste Fourier (1768–1830), French physicist and mathematician, began life as the orphaned son of a tailor. His low social standing prevented him from joining the artillery corps, so he turned his attention to mathematics and was eventually appointed to the Chair of Analysis at the École Polytechnique in Paris. In 1798 he accompanied Napoleon to Egypt and for a time was the governor of Lower Egypt. He is best known for the analytical method that bears his name. By means of Fourier analysis, mathematical functions may be represented, to any desired degree of accuracy, by series of sine and cosine functions – the longer the series, the greater the accuracy. Fourier's interest in physics, especially the temperature of the Earth, led him to propose, correctly, that the atmosphere acts to increase the Earth's temperature. He recognized that air is more transparent to the sun's rays than to the "non-luminous heat" radiated by the Earth.

   The instruments of Fourier's day were not sufficient for him to progress much further in this concept, so it wasn't until decades later that the Irish physicist John Tyndall (1820–1893) began to quantify Fourier's concept. Tyndall showed experimentally that water vapor and carbon dioxide absorb infrared radiation far more strongly than do the oxygen and nitrogen in our atmosphere. This is why temperature usually drops faster on clear nights than on cloudy ones and why water vapor can freeze as frost on grass during clear nights, even when the temperature of the air a meter or two (a few feet) over the ground is above freezing. Radiation cooling can produce a big drop in air temperature near the ground, if the air is still.

The fact that the absorption of infrared radiation by water vapor and carbon dioxide strongly affects different regions of the long-wave infrared spectrum was not known in Tyndall's day. Decades later the Swedish scientist Svante Arrhenius (1859–1927) suggested that a higher level of CO_2 in the atmosphere could raise the temperature of the Earth by absorbing a portion of the long-wave infrared radiation emitted by the Earth, preventing part of this radiation from escaping into space. He viewed the resulting global warming as a good thing that would very likely make the Swedish climate milder.

Arrhenius' suggestion at first elicited an enthusiastic response because of his scientific renown and the interest in ice ages, which was very high at that time. In those days, however, measurements of the infrared absorption spectra of H_2O and CO_2 seemed to show that these spectra completely overlapped each other. Because there is vastly more H_2O than CO_2 in the atmosphere, it was decided that the effect of more carbon dioxide would be very small compared to that of water vapor.

CO₂ and Methane

Not until the 20th century did long-wave infrared measurements become refined enough to show that the infrared absorption maxima of CO_2 and CH_4 actually included different ranges of the infrared than did that of H_2O. As a result, the effect of carbon dioxide and methane is not overwhelmed by water vapor, and both are important absorbers of the Earth's infrared radiation all by themselves. Methane, which molecule for molecule is a much stronger absorber of infrared radiation than is water vapor or carbon dioxide, does not last for extended times in the atmosphere because it is chemically very active. Carbon dioxide lasts a lot longer because it is much more chemically inert than methane. This is why CO_2 is used in fire extinguishers and why methane, the main component of natural gas, is burned to heat homes and offices. Methane reacts in the air with oxygen over times measured in years to produce CO_2 and water vapor. Carbon dioxide is removed from the atmosphere by plants or dissolves into the oceans in times measured in years, decades, and centuries, or even millennia in the case of its reaction with rocks. The study of the life cycle of planetary CO_2 is not a settled matter, but there is no doubt that it lasts a lot longer in the air than methane does.

Carbon dioxide dissolves in water and seawater to form carbonic acid and other carbon-containing chemical species. Such waterborne carbon compounds can be preserved for very long periods of time when they react to form carbonate minerals, such as limestone, or are used by mollusks to form seashells. Water vapor, which is also a greenhouse gas, comes and goes as rain, snow, etc., but is constantly replenished by evaporation. Increasing atmospheric water vapor, carbon dioxide, and methane all act to keep the Earth warm by slowing up the loss of energy as infrared radiation. These gases are called "greenhouse gases" because they increase the temperature of the Earth by preventing some of the Earth's infrared radiation from escaping, in a way somewhat analogous to the glass in a greenhouse blocking infrared radiation from warm plants and soil from escaping but letting in sunlight. Carbon dioxide increases the temperature by transmitting sunlight while at the same time absorbing infrared radiation on its way out from the Earth into space. When CO_2 molecules themselves emit infrared radiation, they do so in all directions, downward toward the Earth as well as upward into space. The infrared radiation they absorb, however, is mostly headed upward. Also, the temperature of the upper air is lower than air near the surface, so the infrared radiation that the upper air emits is reduced in proportion to the fourth power of its temperature, as Boltzmann proved is the case for all things. The warming effect of CO_2 has been calculated in detail, and there can be no doubt about its critical influence on global temperature. Difficulties arise mainly in predicting the exact extent of this warming and what it means for future global weather patterns. After all, the weather predictions for even next week are not all that accurate.

It may be conceptually useful to consider that the CO_2 added to our air by the combustion of fossil fuel yields the same warming (other things like cloud cover staying the same) that would happen if the Earth's orbit were somehow magically moved closer to the sun. If the CO_2 level in the air is considered by itself, and if this CO_2 level doubles from where it is now to about 800 parts per million by volume (ppmv), its warming effect would be equivalent to about a 2% increase in incident sunlight energy, assuming the Earth's average albedo stays constant. Because the rate of increase in CO_2 is itself rising, it may take only about 50 years to double the amount of CO_2 in the air. Doubling present CO_2 levels can increase the Earth's temperature about the same amount as magically moving the Earth's orbit about 1,400,000 miles (2,250,000 kilometers) closer to the sun. Thinking about the global warming effect of CO_2 in terms of moving closer toward the sun gives the global warming effect of CO_2 a physical perspective that may be easier to envision than its effect on infrared absorption.

A very large body of literature has developed concerning global warming. The result of this warming will affect everyone. The melting of non-polar glaciers will increase sea levels, as will the expansion of ocean water as it warms and its density becomes lower. On the other hand, an increase in the amount of ice in central Antarctica will decrease the size of the oceans. Storms, especially hurricanes, which derive their enormous energy from warm ocean water, can become more numerous and more severe as ocean temperatures increase. On the other hand, changing upper atmosphere wind shear might decrease hurricane activity, as it did in the 2006 hurricane season, by cutting the tops off developing storms. Patterns of rainfall may shift and change. Determining what will happen in detail is beyond the limit of our current ability. But it is certain that something will happen, and this something might be very bad.

Most threatening is the potential for runaway effects that lead to uncontrollable global warming. For example, the release of methane from the ground increases when permafrost melts, and rising temperature thaws ever-deeper permafrost layers, releasing more methane. This new methane enhances warming, leading to more methane release, and so on, and so on. Methane also exists in a solid, hydrated form under parts of the seabed. Because solid hydrated methane is actually less dense than seawater, it will float to the surface if dislodged. Once at the surface, and no longer under the pressure of the ocean, it sublimes, forming methane gas plus water. If the temperature of the water in the deep ocean increases, methane hydrates can also decompose under water, and the resulting methane gas will bubble to the surface. Methane plumes have been reported in the East Siberian Sea. If solid methane hydrates begin to decompose, the resulting uncontrolled and massive methane release could be very bad indeed. Some have speculated that sudden increases in atmospheric methane may have contributed to mass extinctions in the past. A meteor impacting the deep seabed could cause the release of massive amounts of methane.

In addition to methane, higher humidity also boosts absorption of infrared radiation by the atmosphere, generating greater warming and more water evaporation, which leads to still more warming. As the oceans warm up, they also evaporate water faster, leading to increased global warming. Warm ocean water holds less carbon dioxide than cold water, as anyone who has opened a carbonated soda bottle warmed in the sun knows. Release of even a small fraction of the CO_2 in the oceans would lead to dramatically increased global temperatures and the release of even more dissolved CO_2. The mass of water in the oceans is more than 250 times the mass of air in the atmosphere. Estimates vary, but there is probably more than 30 times as much carbon stored in ocean water as there is in the atmosphere. If this carbon begins to be released as CO_2 as the oceans warm up, things could get out of hand. Self-amplifying (so-called positive feedback) effects might easily develop beyond human ability to stop them. The planet Venus illustrates just how far out of hand global warming can get. Its dense atmosphere is 96% CO_2, producing a surface temperature so high it can melt lead. That's global warming in spades. The temperature on Venus can be higher than on Mercury, which is about twice as close to the sun, but its atmosphere is very thin and contains only a trace of CO_2.

On Earth, higher CO_2 levels and diminished snowfield and glacier sizes reduce the sunlight reflected back into space and the energy radiated away from the Earth, thereby boosting planetary temperature. On the other hand, some effects induced by global warming might act to decrease warming (so-called negative feedback). Greater snowfall in Antarctica tends to lower the level of the oceans. The increased formation of large, white, puffy water vapor clouds that reflect lots of sunlight could lower the global temperature. Feedback effects, whether positive or negative, can be subtle, complicated, and difficult to calculate, much less control. This is what makes them treacherous.

The Big Picture

Fire, the gift of Prometheus, may indeed prove to be deadly by relentlessly driving upward the CO_2 level in our atmosphere via the ever-increasing combustion of fossil fuel to power the world's growing economy. From the years 1000 to 1800, the level of CO_2 in the air remained relatively constant, at about 280 ppmv (parts per million by volume of dry air). Analysis of air trapped in glacial ice has proven that there is a fundamental correlation between the level of CO_2 and global temperature over hundreds of thousands of years. During the past 1000 millennia, CO_2 varied from just under 200 to about 300 ppmv. At the same time, the average temperature varied by about 18 °F (10 °C), with the low temperatures corresponding very closely with CO_2 levels below 200 ppmv and the highest temperatures corresponding to CO_2 levels of just about 300 ppmv. The question is, did the changing CO_2 level drive the temperature, or did the changing temperature drive the CO_2 level?

One clue to the answer is the cyclic nature of ice ages and the concomitant changes in the Earth's tilt, wobble, and elliptical orbit discussed in the previous chapter. The Earth's geological temperature matches well with the warming–cooling cycles predicted by Croll and Milankovi . The higher CO_2 levels found in ice cores correlate with global temperature increases. It is very likely that these CO_2 increases result from the planetary temperature variations caused by changing planetary tilt, wobble, and orbit effects. Of course, an ice age will affect plant growth, and plants consume CO_2, but dead plants yield most of their stored carbon back to the atmosphere as CO_2 when they rot. If dead vegetation turns into coal, oil, or natural gas, then the carbon in plants is sequestered and stored for eons. In the normal cycle of plant growth, death, and decay, the carbon that plants contain is continually recycled. Seashells, because of their calcium carbonate content, can also sequester carbon dioxide for eons. Carbonic acid in rain interacting with rocks also sequesters carbon for a very long time.

The geological processes that produce fossil fuels take millions of years to develop, far longer than the 100,000-year ice age cycle. A rapidly acting (geologically speaking) carbon cycle that depends on temperature is needed to explain how global CO_2 levels can closely track ice ages. The most obvious answer is the fact that CO_2 dissolves better in cold water than in hot water. The colder seawater that existed during global ice ages would have more capacity to take up CO_2 from the atmosphere than the warmer ocean water that accompanies global warming. Because atmospheric CO_2 can grow by the warming of the oceans as global temperature goes up, what can we expect now that the massive burning of fossil fuels is boosting the level of CO_2 in our atmosphere all by itself and the temperature of the oceans is also rising? Let's look at recent atmospheric CO_2 numbers and see if we can draw some conclusions.

Figure 3.1 shows just how dramatic the current rate of increase in atmospheric carbon dioxide really is.

The annual fluctuation of atmospheric carbon dioxide can be seen from data measured at the monitoring station on top of Mauna Loa. This very evident seasonal variation occurs due to seasonal changes in the Northern Hemisphere, when trees are either growing or dormant. When trees and plants stop growing in winter months, carbon dioxide in the atmosphere increases, and when spring comes it fluctuates downward again. The upward trend is obvious. What about times in the

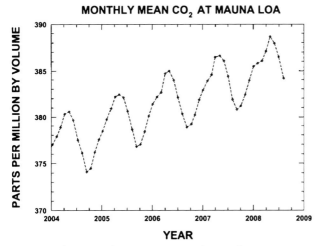

Figure 3.1 The seasonal variation in atmospheric carbon dioxide as recorded on top of Mauna Loa, Hawaii, since 2004. The variation in carbon dioxide level that occurs due to seasonal changes in the growth of vegetation in the Northern Hemisphere is clearly evident. The trend line shows an average increase well above one part per million by volume (ppmv) per year.

past that are much longer than one year? In 2008 the planetary CO_2 level was about 386 ppmv and was increasing at more than 1.4 ppmv per year. Clear and conclusive results from the analysis of trapped air and ice extracted from cores taken from deep inside glaciers show that the level of CO_2 in the air is now about one-third higher than it has been for more than 800,000 years, as shown in Figure 3.2. This figure also shows how the variation in the average temperature closely tracks the CO_2 level.

If it is true that in the past global temperature changes brought on by Earth's orbit, axial tilt, and wobble warmed up the oceans a bit and thereby released some of the CO_2 they contained, what does that mean for us now? Any release of the CO_2 stored in the ocean would have major importance regarding what may happen as the average temperature of the Earth and its oceans is driven upward by the forced increase in atmospheric CO_2 due to the burning of fossil fuels.

The rapid change in CO_2 levels that is occurring now will very likely exert major influence on the planetary heat balance all by itself, even without feedback effects. As the temperature of ocean water increases, CO_2 could be released to the atmosphere from the oceans in amounts far beyond that added by combustion of fossil fuels. The level of atmospheric CO_2 is already increasing at a rate not seen before in the historical record, and this increase is accelerating. Even if the rate of growth could be moderated enough to stabilize the atmospheric level at about 550 ppmv, the resulting temperature rise could be in the range of about 2–5 °C. If the warming oceans begin to yield back to the air even a small fraction of the dissolved CO_2

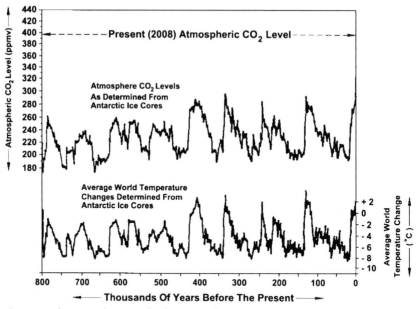

Figure 3.2 The atmospheric CO_2 level over more than 800,000 years as determined from analysis of air bubbles and hydrogen isotopes trapped in Antarctic ice over this same period. From ice core air bubble and isotope data, it is also possible to determine the temperature over this same time period, and temperature closely tracks atmospheric CO_2 level. These data are consistent with the concept that changes in Earth's temperature caused by orbit, tilt, and wobble variations, and accompanying temperature changes in Earth's oceans, result in the release or the uptake of carbon dioxide by ocean water, thereby varying the level of CO_2 in the atmosphere on an approximate 100,000-year cycle.

they contain, or if seabed solid methane hydrates begin to decompose, the resulting increase in global temperature might become uncontrollable.

It is always tricky to fool with Mother Nature. The safest thing to do is to slow down the rate at which human actions are adding CO_2 to the atmosphere. At the current growth rate of world population and industrial activity, the CO_2 level in the air might not stop growing until all economically available fossil fuels are burned. What happens at that point? Just how much of this fuel is there, and how long might it last? When the insatiable need of the industrial world for energy has consumed all the available fossil fuels on the planet, what then will be the level of CO_2 in the air? And how long will this increased level of CO_2 last? No matter what we do, it is certain that the Earth's orbit will continue to change in its 100,000-year cycle from more elliptical to less elliptical, while the tilt of the Earth's poles will precess and wobble. Milanković's calculations show that within 2000 years the resulting changes in sunlight near the polar regions will be sufficient to begin the onset of Earth's slow slide into the next ice age. But by then all fossil fuels may have been consumed, and natural processes will start lowering the Earth's atmospheric CO_2 and methane levels. Global warming might then be a thing of the past. In the long run, in the contest of global warming versus ice ages, ice ages are very likely going to win and glaciers will start returning again, just as they have many times in the past. Is it possible that humanity would have to endure another ice age with only wood fires for heat as we did before? The next chapter looks at the important question of the total quantity of the Earth's supply of fossil fuels and how long these fuels might last.

Recommended Reading

Arrhenius, Svante. "*Nature's Heat Usage.*" Lecture given at Stockholm University, 3 February 1896 and printed in Swedish in Nordisk Tidskrift 14 (1896): 121–130. Translated in *The Legacy of Svante Arrhenius, Understanding the Greenhouse Effect*, eds. Henning Rodhe and Robert Charlson. Stockholm: Royal Swedish Academy of Sciences and Stockholm University, 1998.

Batterbee, Rick and Heather Binney, eds. *Natural Climate Variability and Global Warming*. New York: Wiley-Blackwell, 2008.

Bowen, Mark. *Thin Ice – Unlocking the Secrets of Climate in the World's Highest Mountains*. New York: Henry Holt, 2005.

Houghton, John. *Global Warming: The Complete Briefing*, 3rd ed. Cambridge and New York: Cambridge University Press, 2004.

Lovelock, James. *The Revenge of Gaia: Earth's Climate Crisis and the Fate of Humanity*. New York: Basic Books, 2006.

Solomon, S., et al., eds. *Climate Change 2007: The Physical Science Basis. Contributions of Working Group 1 to the Fourth Assessment Report to the Intergovernmental Panel on Climate Change*. Cambridge and New York: Cambridge University Press, 2007. See also: http://ipcc-wg1.ucar.edu/wg1/wg1-report.html

Weart, Spencer. *The Discovery of Global Warming*. Cambridge, Massachusetts: Harvard University Press, 2003.

4
Earth's Fossil Fuel Supply

*With coal gone, oil gone, high-grade metallic ores gone, no
species however competent can make the long climb from
primitive conditions to high-level technology. This is a one-shot
affair.*

Fred Hoyle, *Of Men and Galaxies*, 1964

Carbon and hydrogen are part of the stuff of life on Earth. We could not exist
without them. Plants typically use about 3% of the energy of the sunlight that
shines on them to combine hydrogen from water with carbon, mostly obtained
from CO_2 in the air, to form a class of chemicals called hydrocarbons. Only a few
plants, such as mushrooms, grow without light by feeding on the organic matter
they extract from soil, and a very few in the deep ocean near volcanic vents manage
to live on sulfurous materials. When plants die, they leave hydrocarbons behind.
As post-mortem decay proceeds, the carbon and hydrogen contained in the hydrocarbons of dead plants are mostly returned to the biosphere as humus, CO_2, and
water. But rotting is not always carried to completion. Occasionally, some of the
carbon and hydrogen compounds from plants are preserved. The removal of CO_2
from air by growing plants, and the preservation underground of the carbon and
hydrogen in their hydrocarbons as fossil fuels, has been going on far longer than
any creatures resembling human beings have been walking around on Earth.
Humanity is now in the process of burning the hydrogen and carbon in the fossil
fuels produced by plants over the last 500 million years.

Coal, petroleum, and natural gas are the three principal fossil fuels. Each can
appear in different forms. Coal, for example, can be anthracite (hard coal), bitu-
minous (soft coal), or lignite (brown coal). Each of these categories can be subdi-
vided further. No matter what its type, coal is the product of organic matter,
compressed underground for millions of years and reduced to different physical
forms of carbonaceous matter by time and pressure. Most geologists believe that
oil and natural gas are also hydrocarbons from long-dead plant matter. But some
have suggested that inorganic chemical processes occurring at extremely deep
levels beneath the Earth can also produce natural gas. After all, there is lots of
methane on Jupiter, and nobody suggests that any of it came from the decay of
plants. The existence of inorganically generated methane on Earth is still under

Energy Demand and Climate Change: Issues and Resolutions. Franklin Hadley Cocks
© 2009 WILEY-VCH Verlag GmbH & Co. KGaA, Weinheim
ISBN: 978-3-527-32446-0

debate. Barring the discovery of large-scale deposits of inorganically produced methane, we can safely assume that the carbon and hydrogen in petroleum and natural gas came originally from plants, using biological processes powered by the sunlight of ancient days.

Like coal, petroleum is mostly carbon but has more hydrogen per carbon atom. As a result, petroleum is a liquid, not a solid. Natural gas, which is mostly methane (CH_4), has even more hydrogen per carbon atom than petroleum and is a vapor, not a liquid, under normal atmospheric pressure and temperature conditions. Because of its high hydrogen content, burning methane produces an exhaust gas with one-third carbon dioxide vapor and two-thirds water vapor (by volume). Burning oil yields an exhaust gas mixture of carbon dioxide and carbon monoxide, as well as oxides of nitrogen, and a small amount of water, which is why water drops can sometimes be seen dripping from the tailpipes of automobiles. Burning coal yields mostly carbon dioxide, ash, soot, carbon monoxide, oxides of nitrogen and sulfur, and some water vapor. It also releases a lot of nasty contaminants, such as selenium and mercury, and various radioactive isotopes, primarily uranium and thorium oxides, accumulated during coal's multimillion-year geological formation process.

Mining coal and moving it to power plants requires considerable energy. Drilling for oil and then refining it also uses energy. Natural gas doesn't need as much processing as petroleum or coal, but compressing it and sending it over long distances via pipelines requires substantial energy. Transporting all of these fuels or electric power from where they are obtained to where they are used has significant energy and financial costs. Transmitting electrical energy over long distances costs the most in terms of energy consumed versus energy transported, and moving oil via ship costs the least. When natural gas is burned, the carbon it contains is converted to carbon dioxide, and the hydrogen it contains becomes water. The exhaust gas from burned natural gas is relatively free of pollutants compared with that from coal or oil, but natural gas does contain very small amounts of the radioactive gas radon. Even so, natural gas offers many advantages, but the combustion of all three fossil fuels – coal, oil, and natural gas – is necessary to keep the world going.

Limits of Fossil Fuels

How much coal, oil, and natural gas exist inside our planet? The answer to this question must be divided into two parts. First, there are fossil fuel *reserves*, that is, the amounts that can be extracted and used with existing technology at current pricing levels. Beyond reserves are fossil fuel *resources*, which are those amounts thought or known to exist that cannot be extracted economically with existing technology. An example of a fossil fuel resource is the petroleum remaining in old oil fields that can no longer be operated profitably even though they still contain oil. Especially for old oil fields, enormous efforts are constantly made to find new ways of profitably tapping their remaining oil. For oil, the extraction limit is reached when the amounts that flow out are no longer worth the cost of obtaining them. For coal, one important limit on profitable mining operations is the minimum thickness that coal seams must have in order to be worked by existing methods. Thicker coal seams are usually more easily worked than thinner ones. For natural gas, the limit is reached when the cost of extracting it becomes higher than its selling price.

With the distinction between *reserves* and *resources* in mind, it is currently true that resources are larger than reserves. It is important to note, however, that resources are converted into reserves as increasing selling price or advancing technology make it profitable to harvest them. At that point, some of the resource base becomes part of the reserve base. If the price becomes high enough and technology advanced enough, all resources can be converted into reserves. Let's look at the total resources and reserves now existing for fossil fuels, beginning with coal.

Coal

Coal is far and away the most plentiful fossil fuel used on our planet. Estimates vary, but there are about one trillion tons (1×10^{12} tons, a shorthand way of writing 1,000,000,000,000 tons) of world coal reserves. Projection of consumption rates a century from now cannot be very accurate, but existing coal reserves should be enough to supply humanity's increasing need for this fossil fuel for perhaps 200 years, if coal's global-warming and pollution effects are ignored or solved somehow. Coal resources are about seven times larger than coal reserves but can be obtained only by increasing the price, via new technology, or both.

Natural Gas

Natural gas is, perhaps surprisingly, Earth's most abundant fossil fuel. This statement is true, however, only if gas resources are included with gas reserves. Like oil, conventional supplies of natural gas (methane, i.e., CH_4) are extracted by drilling holes into underground natural gas deposits and allowing the trapped gas to stream out. The trick is to know where the deposits are. Natural gas deposits are often found in conjunction with oil, and such natural gas is termed "associated" natural gas. Natural gas also exists in pockets that are not associated with oil as well as in dispersed, interstitial gas deposits in combination with coal. The natural gas in coalfields presents a severe hazard to coal miners when it is released during mining operations and forms an explosive mixture with air. Natural gas in coal mines is vented outside of the mine as rapidly as possible and is almost never considered a reserve. However, some coal deposits contain so much associated gas that coal mining is not attempted. Instead, multiple boreholes are used to extract this gas while the coal is left alone. It should not be surprising that different fossil fuels are often found near each other, as they are all products of decayed organic material; however, as noted, some have suggested that natural gas may be created from inorganic materials at great depths below the surface of the Earth. The world's reserves of conventional natural gas, including that obtainable using all known recovery schemes, provide a world energy reserve about half the size of that available from world coal reserves. There is, however, a form of hydrated natural gas that offers even more potential, if it could be harvested economically.

Hydrated Natural Gas

Unlike liquid petroleum, natural gas and water can combine to form a solid gas-water mineral under low-temperature, high-pressure conditions. At the freezing point of water, for example, a solid, hydrated compound of CH_4 and H_2O forms under the pressure of 820 feet (250 meters) of water. There is an amazing amount of this solid, hydrated natural gas in various places in the ocean floor, totaling nearly twice the energy available from the world's total coal supply, reserve and resource coal combined. Unfortunately, no way is presently known to harvest this hydrated natural gas resource economically. When brought to the surface, where the pressure is much lower, solid, hydrated natural gas sublimes, turning directly from a solid into a gas without melting, just like solid carbon dioxide, which is sometimes called "dry ice" because it does not melt and turns directly into vapor under ambient conditions. It looks a lot like ice, but, unlike solid carbon dioxide, the vapor from solid methane hydrate is a combustible mixture of water vapor and methane. This is why hydrated natural gas is sometimes referred to as "burnable ice." Ocean-bed deposits of this solid form of natural gas are typically in the form of small, hydrated natural gas crystals intimately mixed with lots of sand and mud. Because it cannot yet be economically harvested, hydrated natural gas is a vast unused energy resource. Many schemes have been proposed to convert this resource into natural gas reserves by heating the mixture of sand, water, and hydrated natural gas crystals in place on the ocean floor and collecting the resulting vapor before it can convert back to its solid, hydrated methane form. To date, all schemes for harvesting solid, hydrated natural gas have proven impractical.

At the current growth rate of conventional natural gas consumption, this fuel will last about a century, if only natural gas reserves are considered. Counting hydrated, sea-floor methane, natural gas as a fossil fuel would last several centuries, but the cost of natural gas obtained from hydrated natural gas crystals would be far, far higher than gas in pressurized, underground pockets beneath dry land, if it can be obtained economically at all.

Oil

If any fossil fuel can be considered humanity's Achilles heel, it is oil. The world's supply of oil from conventional sources can yield a supply of energy less than one-third that obtainable from conventional coal reserves. Counting all possible oil reserves and resources, whether conventional or not, yields a total oil energy supply roughly one-tenth that available from all the world's coal, reserves and resources combined. The American geologist M. King Hubbert (1903–1989) is well known for correctly predicting that oil production within the United States, not counting Alaska, would reach a peak in the 1970s and decline thereafter. His basic concept was that the total supply of oil is limited, and someday a point will be reached when the amount that can be extracted will begin to decrease. Applying this concept to predict the date when world oil production will start to decline isn't easy, in part because there are not enough data to make an accurate assessment. In addition, new methods for searching for oil deposits are continually being developed, and new areas for exploration sometimes appear, as seems to be the case for the ground beneath the Arctic Ocean, which becomes increasingly available for exploration as Arctic ice recedes. But even so, the world is inevitably approaching this date, whatever it turns out to be, when demand for oil will exceed supply. And that date is not far off, possibly even before the year 2015.

Some have suggested that there is actually not a problem at all with the world's supply of oil and that new sources of oil will always be found as the price of oil rises. It is true that new sources of oil will be developed as the price of oil climbs. Oil sands, for example, have recently been developed because the price per barrel of oil has become high enough to make it worthwhile to mine these sands and heat them to release petroleum. The extraction of oil from shale is being investigated but is still not economical at current oil prices. The energy cost of producing petroleum from oil shale is much higher than producing it from oil sands, and far higher than pumping it from oil wells. Many more tons of oil-shale rock must be quarried than tons of oil sands heated to obtain a barrel of oil. The energy cost as well as the dollar cost of oil from oil shale will be very considerable. The production of oil from oil shale or from coal itself will not reduce the price of oil below its current level but could at least make very expensive oil available. Combining coal and water can also produce oil, but this process produces a lot of carbon dioxide and is environmentally very messy. The energy needed to produce oil from coal and water is considerably higher than just pumping it from holes in the ground.

Currently, high living standards depend in great measure upon low-cost petroleum, natural gas, and coal, and obtaining these fuels from reserves requires less than one-tenth as much energy as that produced when they are burned. As the price of fossil fuels increases, marginal deposits will be harvested but will require an ever-increasing energy input to get to them. The energy gain realized, compared with the energy it takes to obtain them, will decrease as production costs increase. The world's population is growing, but the world's supply of fossil fuels is not. Even as global oil production is still increasing, average oil use per person on our planet already reached its maximum in 1979 and is now steadily decreasing, because population growth is faster than oil production growth.

Sequestration of CO_2

Isn't it possible to keep the carbon dioxide produced by burning fossil fuels from reaching the atmosphere? Many suggest that carbon dioxide can be sequestered by turning it into carbonate minerals, dissolving it in deep ocean water, or pumping it into the ground, and all of this is true. Non-atmospheric, long-term disposal of CO_2, usually termed sequestration, is economically viable now for power stations located near appropriate geological structures such as nearly depleted oil or gas fields. Injection of highly pressurized CO_2 into a depleted field can make additional oil recovery possible. Carbon dioxide serves this purpose in part because its critical point ($31.1\,°C$ and 74 atmospheres) is easily reached in underground gas and oil fields. The critical point is the temperature and pressure at which liquid and vapor are no longer distinguishable. Above the critical point, oil and CO_2 form a solution with a reduced viscosity compared to that of petroleum.

This reduced-viscosity petroleum and carbon dioxide liquid can ooze through underground formations and be extracted more easily than oil itself can. Of course, when brought to the surface, CO_2 becomes a gas again and must be reinserted underground. Injecting CO_2 into oil gas/fields has the advantage of making use of underground geology that retains pressurized gas for geologic times. Injecting CO_2 into the ground and leaving it there is one important way of keeping it out of the atmosphere. But this method is currently viable only if there are oil or gas fields or other suitable underground geological structures near enough to a power station to make it practical.

Special geological structures are needed to keep the CO_2 from oozing back out and reaching the air. Injecting CO_2 into cold, deep ocean water can also be used to sequester carbon dioxide for long periods of time, because very deep ocean water changes places with surface ocean water very slowly. Both of these sequestration methods are not cheap. In general, the cost of the electric power from power plants that carry out sequestration will be higher than that from plants that do not, because the sequestration process uses energy and needs favorable geology that must be close by. Sequestration works only for fixed power stations. Sequestration of the exhaust gases from cars and trucks is virtually impossible. The best way to reduce exhaust gas from fossil-fuel vehicles is to increase the distance they can go using a fixed amount of fuel (see Chapter 10). One way to do this is to decrease vehicle size. Another is to use designs that decrease air drag or stop the engine when the vehicle stops. Large-scale commercialization of high-mileage vehicles will be accelerated by the increasing cost of fuel.

If massive sequestration of the carbon dioxide from the combustion of fossil fuels is not possible, the level of CO_2 in the Earth's atmosphere, after all the available fossil fuel supply has been burned, could become catastrophically high. At our accelerating rate of usage, it seems probable that burning of the Earth's fossil fuel supply will continue until all such fuels are gone, or until more energy is needed to extract them than is gained by burning them. In such a case, truly nightmarish changes in climate will bear down upon the world, driven by incredibly high levels of atmospheric carbon dioxide. Just how high could the carbon dioxide level go?

CO_2 Level Calculations

An estimate is not difficult to make. The total mass of the atmosphere of the Earth has been estimated to be about 5150 trillion metric tons (5150×10^{12} tons). Spread over the surface of the Earth, this atmosphere produces a pressure of about 14.7 pounds per square inch (1 kilogram per square centimeter) at sea level. Coal has an average density of about 80 pounds per cubic foot (1300 kg per cubic meter), and the world's one trillion metric tons of total coal reserves would make a pyramid of coal 300,000 times the volume of the great pyramid of Egypt. Coal consists mostly of carbon, ash, water, and methane, with carbon being the predominant constituent. A reasonable estimate of the carbon in the total world coal reserves (not counting world coal resources, which are about seven times larger) would be 0.8 trillion tons of carbon. Burning one ton of carbon produces 3.67 tons of carbon dioxide, so burning 0.8 trillion tons of carbon produces about 3 trillion tons of carbon dioxide. Adding all this to the atmosphere puts about an additional 580 parts per million by weight (ppmw) or about 880 parts per million by volume (ppmv) of CO_2 into the air. Atmospheric concentrations of carbon dioxide are usually given in parts per million by volume, not parts per million by weight. Because the molecular weight of CO_2 is 44 grams and the average molecular weight of air is about 29 grams, one ppm by weight becomes $1 \times 44/29 = 1.52$ ppm by volume.

Burning all oil and natural gas reserves in addition adds perhaps 25% more CO_2, bringing the total to about 1100 ppmv. Adding this to the approximately 385 ppmv we already have (see Figure 3.1) raises the total carbon dioxide level of the atmosphere to over 1480 ppmv, which is more than four times what we have now. What happens then? The answer very likely depends in part on the carbon cycle, the role that carbon plays in our planet's ecosystem.

The Unending Carbon Cycle

The oceans hold about 30 times more CO_2 than does the present atmosphere. The so-called carbon cycle is, on a grand scale, one way of looking at the interaction between the carbon contained in the oceans, the air, and the land. Some examples here can be informative. We have already talked about the carbon contained in fossil fuels. Burning fossil fuels adds carbon dioxide to the atmosphere. CO_2 in the atmosphere is soluble in water. When it dissolves into pure water, this carbon dioxide mostly forms carbonic acid (H_2CO_3). Carbonic acid is very weak, which is fortunate because it exists to some extent in all the water that we drink. In seawater, carbonic acid interacts in a host of ways to form carbonates of various kinds, including calcium carbonate, which plays a big part in many ocean life forms.

Seashells, for example, contain lots of calcium carbonate. Deposition of the calcium carbonate in seashells into the ocean floor removes the carbon they contain from the biosphere. Over millions of years, this calcium carbonate on the sea floor, as well as carbonaceous minerals, can sometimes be driven into the depths of the Earth by the motion of those large masses of our planet's surface called tectonic plates. Limestone can be created in this process, imprisoning carbon for eons. But at least some of it is returned to the air by volcanic eruptions. All compounds that contain carbon dioxide, calcium carbonate, or sodium bicarbonate decompose on heating and release this carbon dioxide. In the case of sodium bicarbonate, which is also called baking soda, this release of carbon dioxide occurs at around 140 °F (60 °C) and is used to leaven bread. In the case of calcium carbonate, this decomposition occurs at a much higher temperature and produces volcanic eruptions.

Volcanic upheavals induced by the high-temperature decomposition of calcium carbonate are caused by heat diffusing up from the decaying radioactive isotopes contained within the mass of the Earth. This calcium carbonate decomposition process occurring deep beneath the surface produces CO_2 under tremendous pressure, and the resulting explosions can exceed that of nuclear detonations in magnitude.

Other natural processes, such as the growth of plants, also remove CO_2 from the air, at least until the plants rot. The very slow reaction of rocks with the carbonic acid in rain removes CO_2 from the air for much longer. In considering the overall cycle of carbon between the air, the land, and the ocean, it is usually taken as a rule of thumb that 50% of the CO_2 added to the atmosphere is removed by relatively rapid (in geological terms) natural processes.

Natural processes for removing CO_2 from the air, such as dissolution into surface ocean water, operate continuously, and it will take perhaps 200 years to consume all fossil fuel reserves. This fact gives the natural atmospheric purging mechanisms two centuries in which to work, but these natural processes could not remove more than about one-quarter of the CO_2 produced during this period. Even when one-quarter is removed, the remaining atmospheric concentration is still remarkably large, about twice the current level. Major climate changes would ensue. Barring either negative or positive feedback effects, the planetary temperature could rise more than 9 °F (5 °C). And this is only if all present fossil fuel reserves are burned, not fossil fuel resources. In actuality, there is a lot of work being done to convert resources to reserves, and undoubtedly in 200 years a lot of fossil fuel resources would end up being burned, adding even more CO_2 to the atmosphere.

The world's consumption of oil currently exceeds 85 million barrels per day. This is a very large amount indeed, and replacing it with anything else cannot happen overnight. It is very likely that increasing costs for oil, far more than increasing costs for coal or natural gas, will stimulate the production of non-oil-based fuels, such as alcohol and biodiesel, as well as the development of non-traditional sources of oil, such as oil sands, tar sands, and oil-bearing shale. The world's consumption of fuel for vehicles is extremely large. The sooner very serious efforts are made to anticipate and solve the problems associated with oil availability and price and to substitute other fuels for petroleum-based fuels, the better. To accomplish such major changes, technology is critical. Part II of this book includes a survey of existing methods that can be used to produce non-petroleum fuels for powering vehicles (see Chapter 12). But remember that such production is at best difficult and expensive. It may be impossible to produce replacement fuels that are as cheap as gasoline (petrol) once was.

Recommended Reading

Bartlett, Albert. "Thoughts on Long-Term Energy Supplies: Scientists and the Silent Lie." *Physics Today*, 57 (July, 2004): 53–55.

Campbell, Colin J. and Jean H. Laherrére. "The End of Cheap Oil." *Scientific American* (March, 1998): 78.

Carroll, John. *Natural Gas Hydrates*. Amsterdam: Elsevier, 2009.

Deffeyes, Kenneth S. *Hubbert's Peak*. Princeton, NJ: Princeton University Press, 2000.

Hubbert, M. King. *Nuclear Energy and the Fossil Fuels*. Lecture before the Spring Meeting of the Southern District Division of Production, American Petroleum Institute, Plaza Hotel, San Antonio, Texas, March 7–9, 1956. Publication No. 95, Shell Development Company, Exploration and Production Research Division, June 1956. See also: www.hubbertpeak.com/hubbert/

Wilson, Elizabeth and David Gerard, eds. *Carbon Capture and Sequestration*. New York: John Wiley, 2007.

5
Nuclear Power

*There are more things in Heaven and Earth, Horatio,
than are dreamt of in your philosophy.*

Hamlet, Act I, Scene 4

*To smash the little atom, all mankind was intent:
Now any day, the atom may return the compliment.*

Attributed to the German physicist and Nobel Prize
recipient (1954) Max Born (1882–1970)

At this moment in history, humanity is overwhelmingly dependent on three fossil fuels – coal, oil, and natural gas – created in Earth's distant past. When fossil fuels are exhausted, where will our energy come from? The bounty of the Earth includes both fossil fuels and heavy isotopes that can be split to release energy. This latter type of fuel was created not in the distant past as fossil fuels were but in the primordial time before the formation of the sun itself. It consists of very heavy atoms that release energy when they fission, breaking into pieces. The most prominent example of such a natural fissionable material is uranium. The energy available from nuclear fuels is far greater than that from all fossil fuels put together. But there is no free lunch. Obtaining energy from nuclear fuels has its own difficulties.

As with fossil fuels, nuclear fuels can be divided into *reserves* and *resources*. If only current nuclear fuel *reserves* of rich uranium ore are counted, without benefit of advanced technologies such as breeder reactors and nuclear fuel reprocessing (more about this in Chapter 13), the potential for nuclear power obtained from currently mined rich veins of uranium ore reserves is only about 0.3% that of coal reserves. But nuclear *resources* are truly enormous and could produce hundreds of times more energy than burning the total of all coal reserves. Nuclear fuel breeding would increase the energy-producing ability of existing nuclear fuel reserves by a factor of over 100. Where did all this nuclear fuel come from?

Origin of Fuel for Nuclear Fission

Stars like our sun fuse together atoms of light elements (e.g., hydrogen) to produce somewhat heavier elements (e.g., helium). Fusing light atoms produces energy

Energy Demand and Climate Change: Issues and Resolutions. Franklin Hadley Cocks
© 2009 WILEY-VCH Verlag GmbH & Co. KGaA, Weinheim
ISBN: 978-3-527-32446-0

just as breaking heavy atoms apart does (see Chapter 14). Energy is produced by fusing together very light elements such as hydrogen and by breaking apart very heavy ones such as uranium. But stars that operate using the same nuclear reactions as our sun cannot produce uranium atoms. It takes the nuclear cascade chain of a colossal supernova stellar explosion to produce heavy elements. Our sun is mostly hydrogen and helium, but it also contains heavy elements such as iron. It is therefore a second-generation star, formed from the detritus of an earlier supernova explosion. The entire solar system, including the Earth, consists of condensed matter from a stupendous supernova in the primordial past. To echo the American astronomer Carl Sagan (1934–1996), we are all, quite literally, made of stardust.

An atomic nucleus at the center of an atom is tiny, about 100,000 times smaller in dimension than the electron cloud that surrounds it. An atom is mostly empty space, more so even than the solar system we inhabit. But each tiny nucleus carries a positive electrical charge that holds its surrounding negatively charged electrons by electrostatic attraction. Interestingly, it turns out that electrons themselves are so small that their size has never been determined, and it has been speculated that they may not have any size at all, even though they do have a finite mass. It is customary to calculate a size for electrons by assuming that their mass comes entirely from electrostatic energy. This calculated size is about three millionths of a nanometer. If it is true that electrons have no measurable size, it wouldn't be the only incredible thing about nuclear physics.

When heavy nuclei like those of uranium fission by breaking into smaller, lighter nuclei, they release some of the energy used to produce them in the original supernova explosion. In only a few cases can the energy in heavy nuclei be released easily enough to be useful. Producing energy by nuclear fission has its own special problems, not the least of which is the generation of radioactive byproducts. But humanity will increasingly need energy from nuclear sources as fossil fuels become exhausted. How much energy can be produced from the fission of nuclear fuel?

The Energy in Nuclear Fuel

The energy released when a very heavy atom such as uranium splits into smaller nuclear pieces is millions of times larger than when a carbon atom reacts with two oxygen atoms to produce carbon dioxide. When one kilogram (2.2 pounds) of uranium atoms fission into smaller nuclear fragments, the energy released is about the same as that from burning three million kilograms (6.6 million pounds) of coal. What causes this huge difference in energy? The answer is that nuclear forces are millions of times stronger than the electrostatic forces that release energy when coal is burned. The process of burning coal changes the electron bonds in carbon and oxygen. The bound electrons in solid carbon, like those in gaseous oxygen molecules, occupy specific energy levels that are set by the intricacies of quantum mechanics. Burning coal releases energy because the bound electrons in the carbon and oxygen move to lower energy levels in CO_2 than they occupied in solid carbon and gaseous oxygen. If they didn't, coal would be just another incombustible rock. The energy released by the changed energy levels of electrons as they form CO_2 keeps a coal fire burning.

When heavy atomic nuclei are broken into pieces, the energy released is a result of changing *nuclear,* as opposed to *electrostatic,* energy levels. All heavy nuclei contain many protons, and they all should repel each other electrostatically, as they have the same positive charge. Opposite electrical charges attract, and similar charges repel each other. The electrostatic attraction between a nucleus and its electrons is the same type of force that makes light pieces of plastic wrap cling annoyingly to clothing.

The charge on a proton is exactly equal in magnitude to the charge on an electron but is opposite in type (the charge on a proton is positive and that on an electron is negative). Nuclear forces operate only over small distances, but they are so strong that they can hold protons together in atomic nuclei against electrostatic repulsion force that would otherwise make these protons fly apart. The force that holds the nucleus together is entirely different and far stronger than the electrostatic attraction between the nucleus, which carries a positive charge, and its electron cloud, which carries a negative charge. Nuclear forces are far stronger than electrostatic forces, and this is why breaking apart all the atoms in one kilogram of uranium yields more energy than burning three million kilograms of coal.

A uranium nucleus has 92 protons. They should repel each other and fly apart due to electrostatic repulsion, but they don't. What holds them together? The answer is the *strong nuclear force.* This strong nuclear force causes an attraction between protons that counteracts the electrostatic repulsion trying to force them apart. The energy levels in nuclei change as heavy atoms split apart, and it is these changes that release energy in operating nuclear reactors. These changes in energy levels are the nuclear equivalent of the electron energy level changes that occur when carbon burns.

It turns out that this strong nuclear force acts only when protons get extremely close together in the presence of at least one neutron. A neutron has no electrical charge. Without neutrons, the only atoms that could exist would be hydrogen atoms. When a heavy atom, with lots of protons (of higher atomic number than iron with 26 protons, cobalt with 27 protons, or nickel with 28 protons) breaks apart, the resulting changes in nuclear energy levels release energy. Because the nuclear force is extremely powerful, the energy released when a heavy atom breaks apart is large enough to result in a measurable decrease in mass. This fact means that it takes much less nuclear fuel than coal to produce a large amount of energy.

Nuclear Energy

Albert Einstein (1879–1955) famously pointed out that mass may be converted into energy ($E = mc^2$) and vice versa. Because the speed of light, c, is about 186,000 miles (300,000 kilometers) per second, very small changes in mass can produce a lot of energy. Likewise, it takes a lot of energy to produce even a trivial amount of mass. When energy levels in a nucleus change, these energy level differences translate via Einstein's equation into a change in mass. The root source of nuclear energy is the difference in mass that occurs when heavy atoms fission into smaller pieces and nuclear energy levels are altered. When the carbon in coal burns, the changes in the electron energy levels of carbon atoms is minuscule compared with the changes in nuclear energy levels when uranium atoms split apart.

The change in mass that occurs when carbon combines with oxygen must not be confused with the fact that burning coal combines the mass of oxygen atoms with that of carbon atoms. The mass of the resulting exhaust gas is a lot more than the mass of the carbon that was burned. But the combined mass of all the oxygen and carbon atoms before combustion is more than the mass of all the resulting CO_2 molecules after combustion. Measuring the amount of atomic mass lost when coal is burned with oxygen is virtually impossible, in part because the products of combustion are gaseous and occupy an enormous volume and also because the loss in mass per molecule of carbon dioxide is extremely small. But in power plants burning coal or fissioning uranium atoms, both plants running at the same efficiency, the change in mass that powers a typical 1000-megawatt plant for one year amounts to about 115 grams, which is a bit more than four ounces. The change in mass for each of the atoms that take part in a chemical reaction is tiny compared to the change in mass that occurs for each nucleus that fissions. It is not customary to talk about changes in mass when chemical reactions happen, but it is nonetheless true that minuscule mass changes do occur when carbon is burned. Given that there is a lot of energy locked up in nuclear fuel, how much nuclear fuel is there? To answer this question, we must talk about isotopes.

Isotopes

There is nothing mysterious about isotopes. All atoms are isotopes. The term itself derives from the Greek word for "same place" and is an appropriate name for atoms of an element with different atomic weights but the same atomic number. Atoms with the same atomic number occupy the same position in the periodic table of the elements proposed by the great Russian chemist Dmitry Mendeleyev (1834–1907) and universally used since his time. The number of protons in its nucleus determines an element's atomic number and its position in Mendeleyev's table. All atoms of uranium (U), for example, have 92 protons; hence, the atomic number of uranium is 92. All hydrogen (H) atoms have only one proton, so the atomic number of hydrogen is 1. Different isotopes of the same element can contain different numbers of neutrons. The isotope uranium-235 contains 92 protons and $(235 - 92) = 143$ neutrons. Because all uranium atoms have 92 protons, this notation can be simplified to uranium-235, without mentioning the number of neutrons. What is the origin of the strong nuclear force that keeps these 92 protons held together instead of flying apart? That's where neutrons come in. If a proton approaches a neutron, there is no repulsion between them because a neutron has no electrical charge. The wonders of subatomic physics begin when a neutron comes close enough to a proton to exchange a smaller particle called a pion (a short way of saying pi-meson), and this pion exchange produces the strong nuclear force. It was the Nobel Prize–winning American theoretical physicist Richard Feynman (1918–1988) who developed a diagrammatic approach (called Feynman diagrams) for describing the behavior of interacting atomic particles, including pions and the strong nuclear force.

Pions themselves are one of the very few nuclear particles to be predicted before being discovered experimentally. The Japanese physicist Hideki Yukawa (1907–1981) predicted them in 1935, two years before their existence was demonstrated. He was awarded the 1947 Nobel Prize in physics for this achievement. Because of pions and neutrons, atoms can contain even more than 100 protons, although all those with more than 83 protons (the element bismuth) are unstable and radioactive. Inside all atomic nuclei except simple hydrogen, which has only one neutron, pions pass rapidly back and forth between protons and neutrons. When a neutron captures a pion, it becomes a proton, and when a proton loses a pion, it becomes a neutron. Back and forth they go, trillions of times per second, in the unimaginably rapid dance that produces the strong nuclear force holding protons together in atomic nuclei. Feynman's work helped explain how this rapid exchange is equivalent to an attractive force. There are indeed more things in Heaven and Earth than were dreamt of in Horatio's philosophy. But even nuclear power has limits, as it takes nuclear fuel to run a nuclear reactor.

Limits of Nuclear Fuel

Currently, only veins of ore that contain about 500 parts per million (ppm) or more of uranium are mined commercially. Worldwide, such rich vein deposits of uranium ore could yield about 100 million kilograms of uranium, of which 0.7% (730,000 kg) is uranium-235. However, as the grade of ore drops to 50 ppm of uranium, the total amount of ore available increases by a factor of 300. The energy from our planet's total supply of uranium far surpasses that from the world's coal resources even if only uranium-235 is used. As we will see in Chapter 13, it is possible to produce new, fissionable nuclear fuel from uranium-238, and there is more than 100 times as much uranium-238 as there is uranium-235 on our planet.

Extremely low-grade uranium ores, such as granite, are in vast supply. But extracting uranium from granite is extremely expensive in terms of both money and energy. Producing uranium from granite requires not only mining large volumes of rock but also crushing this rock. To get a sense of the energy cost of extracting uranium from granite containing only one part per million of uranium, consider that crushing this ore to one-micron-size pieces to extract the uranium requires more energy than can be obtained from the nuclear fission of the uranium-235 it contains. But pegmatite and sandstone ores are much richer in uranium than granite and might become useful sources of uranium if the cost of energy climbs high enough. The energy and economic costs of extracting uranium from low-grade ore will be higher than those of mining rich uranium ores, but this nuclear energy source would at least be available.

Uranium is also present in seawater, but only at trace levels. Even though the uranium content of seawater is extremely low, amounting to about three milligrams per cubic meter, the energy cost of extracting it is also low because seawater, unlike granite, does not need to be crushed. Instead, uranium can be extracted using ion-exchange resins that preferentially absorb uranium from the water. This process is slow but does not require much energy. Small amounts of uranium have already been collected from seawater in this way. There is 50,000 times more uranium available in seawater than from veins of uranium ore. That's a lot of uranium in the Earth. But if only rich uranium ores are used, then Earth's uranium can fuel nuclear reactors for less than 50 years. If lower-grade ores and resources are used, even without reprocessing or breeding, there is enough uranium-235 alone to provide 1500 times the energy available from all coal reserves. If fuel rod reprocessing and breeding of uranium-238 are carried out, the uranium in seawater alone could generate more than 150,000 times the energy in the world's coal reserves. Of course, all the uranium in seawater cannot be extracted, but if only 1% could be harvested, the energy it could produce still dwarfs that in all coal reserves and resources combined. Breeding would also enable thorium, a so-called fertile isotope, to be converted into a nuclear fuel (see Chapter 13), and there is three times as much thorium as uranium. The question is, how is the energy stored in uranium released in a controlled way to generate electric power?

The Basics of Nuclear Fission

When uranium-235 splits (fissions) apart, it forms the very unstable atom U-236, which rapidly splits into smaller atoms and one, two, or three neutrons, releasing lots of energy in the process. This energy is released as neutrons and fission fragments, such as barium and krypton, all moving at high speed, and high-energy gamma rays, as shown in Figure 5.1.

Without uranium-235, nuclear energy would be only a physicist's dream, not an engineer's challenge. Uranium-235 is the only naturally occurring isotope that can be used to start a chain reaction, in part because it is virtually the only one of the first 92 elements that produces multiple neutrons as it spontaneously fissions. Protactinium-231 with 91 protons and plutonium-239 with 93 protons are two other isotopes that fission and release multiple neutrons, but there are only the tiniest minute traces of natural protactinium or natural plutonium around. Of course, lots of plutonium has now been produced in reactors. But the nuclear age could not have started with plutonium because it was so rare until we started making it. There are lots of heavy atoms naturally available on Earth. However, only uranium-235 can be used as-is to produce power. Without two or more

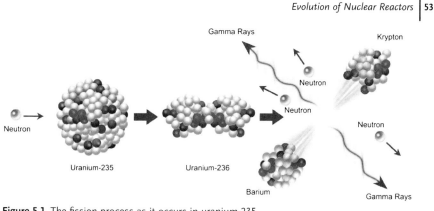

Figure 5.1 The fission process as it occurs in uranium-235. Because multiple neutrons are emitted when uranium-235 is caused to fission by a single neutron, it is possible to initiate a fission chain reaction as other uranium-235 atoms each absorb a neutron and then themselves undergo fission, releasing still more neutrons.

neutrons being produced each time an atom fissions, a chain reaction of millions of rapidly fissioning atoms can't get started. Multiple neutrons are necessary to produce lots of fissioning atoms, as it takes only a single neutron to split it apart and release multiple neutrons.

The release of multiple neutrons when uranium-235 atoms split causes a chain reaction as more uranium-235 atoms are hit by these neutrons, causing these atoms to also fission and release still more neutrons. It is this chain reaction that makes nuclear power possible. The total mass of the fragments produced by fissioning of uranium-235 is always less than that of the original uranium-235. This difference in mass is converted into energy. Current nuclear power reactors "burn" (split apart by fissioning) only a portion of the uranium-235 in their nuclear fuel because, unlike a coal fire, the nuclear fission chain reaction effectively ceases before its fuel is all gone.

Evolution of Nuclear Reactors

In the nuclear assembly line of a supernova, both the uranium isotopes uranium-238 and uranium-235 are created, but uranium-235 undergoes radioactive decay faster than uranium-238 does. At the Earth's present age only 0.7% of natural uranium is uranium-235; virtually all the rest is uranium-238. Earlier in our planet's history, a greater percentage of uranium was the uranium-235 variety, making it possible for natural nuclear reactors to be created by the geological processes that concentrate uranium ore. What remains of such a natural nuclear reactor was discovered at Oklo, in Gibbon, West Africa, in 1972. At that location the isotope mixture found in a uranium ore deposit proved that a nuclear reactor had operated there about 1.7 billion years ago. Such a natural nuclear reactor was possible at that time because of the different half-lives of these two isotopes. As the name implies, half-life is defined as the time required for one-half of the total initial amount of a radioactive isotope to decay. The half-life of uranium-235 is 710 million years, and the half-life of uranium-238 is 4.51 billion years. Calculation reveals that 1.7 billion years ago there was about 3% uranium-235 in natural uranium, and that is about the level of uranium-235 enrichment in most nuclear reactors. Enrichment is the difficult process of boosting the uranium-235 content of natural uranium so that it is higher than the 0.7% level present in natural uranium.

It is worth noting that an isotope, such as uranium-235, decays only when something unusual happens inside its nucleus. Remember those pions and their wild exchange dance? There are a lot of seconds in 1.7 billion years, and in every one of them, pions changed partners trillions of times. Whatever chaotic process causes a particular atom to suddenly undergo radioactive decay after the pions within it have danced for a billion years is hard to imagine. Horatio, you didn't even know the half of it, and neither, perhaps, do we.

Reactors that use so-called heavy water made with deuterium (hydrogen that has one proton and one neutron) can operate with the 0.7% of uranium-235 present in unenriched uranium. Reactors using uranium that has not had its uranium-235 content increased are possible because the hydrogen in heavy water already has a neutron and thus does not absorb another neutron as easily as the hydrogen in regular water does. Reactors that use normal, so-called light water lose some of their neutron flux in this normal water and need uranium enriched to contain around 3% of uranium-235, which is about the level of this isotope in nature's ancient Oklo reactor.

Present-day Nuclear Reactors and Power Plants

Most reactors don't use heavy water and natural uranium. Instead they use normal water and enriched uranium. As it happens, the first nuclear reactors for the production of power were engineered for propelling nuclear submarines. Submarines are not all that large as warships go, and reactors designed for them are compact. From the engineering viewpoint, this means that they have to use highly enriched uranium rather than natural uranium. The fuel rods for nuclear submarines use highly enriched uranium, which is enormously expensive but lasts a lot longer before it needs replacement. In both military and civilian nuclear power reactors, it is heat from the nuclear fission process that boils water and generates the steam that drives turbines and provides power, just as steam in fossil fuel power plants does. The water used to cool most power reactors is pressurized to keep it from boiling. High pressure allows temperatures higher than the normal atmospheric pressure boiling point of water to be attained, just like it does in a pressure cooker. This pressurized water does not boil as it passes over the hot fuel rods, although some so-called boiling water reactors are also in operation. The cycle of water flow in a standard pressurized-water nuclear reactor is shown in Figure 5.2.

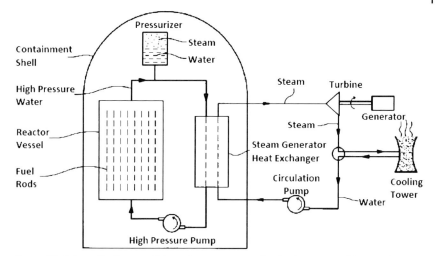

Figure 5.2 Schematic diagram of the water and steam flows in a typical pressurized-water nuclear power plant. The reactor is entirely within the containment shell. The water used to cool the fuel elements in the reactor core is highly pressurized to about 2250 pounds per square inch (15.5 megapascals) to prevent it from boiling. Because it passes repeatedly through the reactor, this pressurized water becomes radioactive, primarily from the hydrogen isotope tritium (hydrogen with two neutrons), formed by the interaction of the neutron flux in the reactor with normal hydrogen (which has no neutrons), as well as from the hydrogen isotope deuterium (hydrogen with one neutron). This highly pressurized water passes through a heat exchanger, where it boils water to produce the steam that is used to spin a turbine and generate electricity. After exiting the turbine, this steam is condensed back to liquid water, usually in a cooling tower, and then pumped back into the heat exchanger, where the cycle begins anew.

As the caption for Figure 5.2 describes, the water used to cool the hot fuel rods and remove the heat released by fissioning atoms is passed through a heat exchanger, where it boils water and produces the steam needed to drive a turbine and generate electricity. As a reactor operates, it generates an increasing amount of the fissionable plutonium isotope plutonium-239 from the reaction of uranium-238 atoms with neutrons. Plutonium-239 is fissionable, just as uranium-235 is. It is far more radioactive than uranium-235 and has a half-life of only about 24,000 years. That is why there is virtually none of it left around from that primordial supernova of long ago. Over time, an operating reactor produces an increasing proportion of its total energy from the fissioning of plutonium-239 and less from fissioning uranium-235. In the first few days and weeks a reactor runs, the plutonium it produces can be chemically extracted and used as-is to produce atomic bombs. After a year of operation, the plutonium mixture is so poisoned with other plutonium isotopes, especially the non-fissionable isotope plutonium-238, that it can no longer be made into nuclear weapons without enrichment. Fuel rods, in a reactor that uses graphite rather than water to slow the neutrons down so they can react easily with uranium, can be removed readily for the extraction of plutonium without shutting the reactor down. Unfortunately, graphite can catch fire and burn just as coal does. A burning nuclear reactor is a nasty thing, as the 1957 fire in the British Windscale reactor and the 1986 fire in the huge Russian reactor at Chernobyl eloquently testify. Both of these reactors used graphite instead of water to slow down the neutrons produced by fission.

Used Fuel Rods

Major waste products from fossil fuel power plants are carbon dioxide, ash, and sulfur dioxide. The major waste products from nuclear power plants are used fuel rods. Used fuel rods are extremely radioactive, especially initially, and contain unburned uranium and plutonium as well as lots of other radioactive isotopes. In electricity-generating power reactors, fuel rods must be replaced after two to three years of operation. In reactors that use natural uranium, a smaller fraction of the uranium-235 can be consumed before replacement is required. Why is this so? Because the level of the fissionable material drops below the level needed to maintain power production. That condition happens sooner for reactors using natural uranium than for those using enriched uranium. As a reactor operates, fuel rods swell in part due to the generation of new elements, some of which are gases. Swelling is also caused by the generation of defects, like vacant spots such as vacancies and voids, which are produced by neutrons knocking the atoms in the rods around. Under normal conditions a nuclear power plant does release some radioactive isotopes, primarily as radioactive gas such as tritium or water vapor containing tritium, into the environment. The radioactivity released in this way is actually about the same as that released by a coal-fired power plant of the same size. The half-life of tritium from nuclear power plants is far, far lower than the half-lives of the isotopes released by coal-fired power plants, and so it is virtually eliminated from the environment in a few decades. Both radioactive uranium and thorium are present at the parts–per-million level in coal, and burning coal releases small amounts of these substances into the biosphere, where they last for eons.

As currently used in commercial power reactors, only a few percent of the uranium-235 is "burned," because of swelling and distortion of fuel rods. One of these reaction products is plutonium, which is itself fissionable. As the reactor operates, plutonium accumulates in fuel rods. After a few months of operation, a significant fraction of reactor power comes from the fissioning of plutonium rather than uranium. Replacement of used fuel rods is usually done partially, by adding new rods to the outside of the core, just as fresh logs are added to a fire, and removing those from the core center in stages spaced about a year apart. The neutron flux density is higher at the center than at the edges of the core, and moving rods around in this way maximizes the burning of the fuel they contain.

The unburned fuel that remains in used fuel rods can be recovered and used to make new fuel rods, but reprocessing is not a simple business. A major investment is required to construct a nuclear reprocessing facility. As rich sources of natural uranium ore are consumed, it is very likely that reprocessing will become a worthwhile thing to do. Meanwhile, used fuel rods need to be stored. The volume of used fuel rods from a 1000-megawatt power plant is not large compared to the thousands of cubic meters of ash generated by coal-fired plants. The problem of storing used fuel rods is solved if they are reprocessed to extract the uranium and plutonium they contain. The reprocessing of fuel rods is very limited currently, so most are just stored. Even if reprocessing is carried out, there are still substantial amounts of highly radioactive materials left over that must be stored for hundreds of thousands of years, and for some isotopes, millions of years. Storage of this waste is a major problem, but at least it does not contribute much to global warming. This radioactive waste is very dangerous and will kill those who get too close to it for very long. The question of the health effects that radiation can produce is not simple. After all, ionizing radiation is used routinely in a large number of medical situations, such as looking for cavities in teeth and treating cancer.

Radiation, Radioactivity, and Health

Life on Earth has evolved over billions of years in the Earth's low-level background radiation from cosmic rays and from the air and soil. As a result, the deoxyribonucleic acid (DNA) that is part of all living things has developed some ability to withstand radiation. If it hadn't, the rate of genetic mutations would be much higher than it is.

DNA, the double helix of life, has evolved groups of molecules that can, in effect, repair or prevent some radiation and other damage to chromosomes, which are single, large, macromolecular pieces of DNA. As a result, animals and plants seem to have some resistance to injury from radiation, but high radiation levels cause problems. Exposure to high levels of ionizing radiation for short times or somewhat lower levels for longer times can lead to serious health problems, especially cancer, and even death. This is why radioactive waste entering the environment and the use of certain x-ray diagnostic procedures, such as computer-aided tomography (CAT) scans, should be considered carefully.

Radioactivity was initially discovered as a property of certain minerals, especially uranium ores. Everyone living on the earth is exposed day and night to a background flux of natural radiation. Charged particles from the sun and the galaxy are the main radiation in space. On the Earth, the main source of radiation is radon, a radioactive gas from the decay of radium. It is present in natural gas. Radon delivers about 55% of the typical average annual radiation dose received by people worldwide, which is about 2.4 millisieverts (see Appendix II for an explanation of radiation dosage units), equivalent to 24 modern chest x-rays. Only about 8% of background radiation on Earth comes from cosmic rays. Computer-aided tomography x-ray scans (CAT scans) deliver a radiation dose of about 10 millisieverts, roughly as much as 100 standard chest x-rays, which deliver about 0.1 millisieverts each. Dental x-rays deliver smaller radiation doses than do chest x-rays, and far less than CAT scans. For comparison, a whole-body dose of about

5000 millisieverts produces a 50% mortality rate. On the other hand, radiation can also be used to destroy cancerous tumors. If enough ionizing energy is delivered to tissue, that tissue dies. The problem is to kill cancerous tissue without harming normal tissue, and many methods are used to accomplish this. One way is to irradiate a tumor from multiple directions, thereby limiting the radiation delivered to any given volume of normal tissue.

Besides radon and cosmic rays, radiation also comes from rocks and soil, including the background radioactive materials we eat in foods. Plants, for example, take up radioactive isotopes that exist naturally in the ground. They also take up radioactive fallout from atmospheric testing of nuclear bombs, and contamination from past atmospheric tests continues to exist in milk and other foods to this day. Cows eating grass and plants grown on soil containing the radioactive isotope strontium-90, which was left over from aboveground bomb tests, incorporate a small amount of this strontium into their milk. There were hundreds of above ground tests carried out in the period 1945–1963, and as a result radioactive strontium-90 can be detected worldwide. Strontium is chemically similar to calcium and ends up in bones and teeth. Fortunately, the half-life of strontium-90 is only 28.5 years, and the Nuclear Test Ban Treaty, signed in 1963, was intended to eliminate atmospheric testing of nuclear weapons, although some nations have conducted aboveground nuclear tests since then. Whether or not this strontium-90 has contributed to an increased rate of childhood leukemia is a controversial question. Mercifully, the level of radioactive strontium in milk steadily decreases as time goes by. More than half of it is gone already. Despite the effect of radioactive fallout, most background radiation, whether from food, soil, or air, is of natural origin. Most radioactive waste lasts a lot longer than strontium-90, and safe storage of this waste over geologic times is a very difficult problem indeed.

Natural Radiation and Radioactive Waste

The radiation level on our world would be much higher if it were not for the fact that the Earth's atmosphere and magnetic field provide shielding against radiation from space. On the moon, for example, which has virtually no atmosphere or magnetic field, the radiation level is typically about 60 times higher than that on Earth. Some of this radiation arrives in the form of cosmic rays, which are not actually rays but high-speed atomic particles. The sun also emits lots of high-speed particles, mostly protons. These protons cause auroral lights as they spiral around the magnetic field lines of the Earth until hitting the upper atmosphere high above the magnetic poles, causing the upper air to glow. The weak but gigantic magnetic field of the sun in the inner solar system also provides some shielding against radiation from the cosmos, and the cosmic ray intensity on Earth goes up and down measurably as the magnetic field of the sun waxes and wanes over the 11-year solar cycle.

Radioactive waste is inevitably a byproduct of nuclear reactors. Used fuel rods are extremely radioactive, especially when first removed, but their radioactivity diminishes rapidly at first and then decreases more slowly. Their rate of radioactive decay changes because the various radioactive isotopes contained in used fuel rods decay with half-lives ranging from less than an hour to millions of years. Remember that the half-life is the time required for one-half of the total amount of any radioactive isotope to decay. Used fuel rods remain dangerously radioactive for a very long time. Large amounts of radioactive waste are also left over from the production of nuclear weapons. Some of this waste from nuclear weapon production was thrown into the ocean and is now unrecoverable. What to do with radioactive waste is an unsettled question, but what appears most practical is storage under dry land, not in the ocean or Antarctic ice. Incredibly, it was once seriously suggested that disposal of high-level waste could be achieved by depositing it on the Antarctic ice sheet, as it would produce enough heat to sink itself into the icy depths, which it indeed would. Fortunately, this disposal concept was abandoned before it could be implemented.

Radioactive isotopes decay at different rates. Some can last for geologic times. The ones that last the longest tend to be the least radioactive. Conversely, those that decay rapidly are highly radioactive. The radiation produced by nuclear waste appears as either electromagnetic radiation such as x-rays or gamma rays, charged nuclear particles such as electrons or protons, and uncharged particles such as neutrons or neutrinos. All of these can cause radiation damage, but the least harmful by far are neutrinos, which hardly interact with anything at all. An average neutrino could pass through a piece of lead about a light year thick with no effect. Neutrinos have no measurable effect on living things. The other particles, as well as electromagnetic radiation, are absorbed to a greater or a lesser degree in tissue and other materials and deliver different amounts of ionizing energy (see Appendix II). This radiation can change the electrostatic bonds in whatever material they hit, and this is why long term storage is necessary.

Disposal and Storage of Nuclear Waste

Currently, burial in very dry, uninhabited areas seems to be the preferred long-term disposal and storage method. Some radioactive waste is soluble in water. To slow down exposure to the environment, encapsulation in massive solid glass is widely used, but glass itself is affected by radiation. A clear glass ashtray dangled overnight on top of used fuel elements stored in a water tank will turn black as a result of the radiation damage the glass undergoes. Glass is also slightly soluble in water. A more permanent method of preventing release of the radioactive isotopes of elements is their incorporation into synthetic minerals related to those that contain radioactive elements in nature. This concept was first suggested by the Australian geophysicist and geochemist A. E. Ringwood (1930–1993), who proposed using special forms of magnesium iron silicates to store radioactive waste. Such minerals are far less soluble in water than glass is, but storing radioactive waste in synthetic minerals is very expensive compared with storing them in glass. In the meantime, most radioactive waste is stored in a variety of ways, including just keeping it in steel containers. Thousands of these containers are just old oil barrels and are rusting and beginning to leak.

The long-term storage of radioactive waste can be achieved technically, but where and how to store nuclear waste are far from settled. As of 1 January 2007, world production of electric power from nuclear reactors was about 372,000 megawatts, the equivalent of 372 typical coal-fired power plants each producing 1000 megawatts. It has been estimated that the total supply of electricity from nuclear reactors will increase to 509,000–663,000 megawatts by 2030. Keeping these reactors supplied with fuel will, sooner or later, require reprocessing to recover the unburned uranium-235 and the fissionable isotope plutonium-239 that used fuel rods contain, and this process will produce nuclear waste that must be stored. It is also possible to breed fresh nuclear fuel, as discussed in Chapter 13, and thereby increase the availability of nuclear energy more than one hundred–fold. We will run out of coal a lot sooner than we run out of nuclear fuel.

As more nuclear sources of energy are used, there will be an increasing need to develop means and methods for the safe, very long-term storage of radioactive waste. Encapsulation in a stable matrix mineral, combined with storing the resulting encapsulated waste in dry, geologically stable depositories, may be the best long-term storage method. A lot of science and engineering, as well as politics, will be needed before the nuclear waste disposal problem is solved. Relentlessly increasing world demand for energy makes long-term storage of nuclear waste inevitable, barring the collapse of civilization. Nuclear sources of energy are going to be used and developed increasingly. Nuclear energy is not the only way to go, however, and Part II presents the potential and the limitations of non-nuclear–, non-fossil–based renewable sources of energy and power.

Recommended Reading

Deffeyes, Kenneth S. *Beyond Oil: The View From Hubbert's Peak.* New York: Hill and Wang, 2005.

Hore-Lacy, Ian. *Nuclear Energy in the 21st Century.* Amsterdam: Elsevier, 2007.

Lewis, Elmer E. *Fundamentals of Nuclear Reactor Physics.* Amsterdam: Elsevier, 2008.

Nuclear Energy Institute: www.nei.org/resourcesandstats/. (This Web address connects to the Nuclear Energy Institute's large collection of data and statistics on nuclear power generation.)

Part II Answers

Energy Demand and Climate Change: Issues and Resolutions. Franklin Hadley Cocks
© 2009 WILEY-VCH Verlag GmbH & Co. KGaA, Weinheim
ISBN: 978-3-527-32446-0

Introduction

If they make you not then the better answer, you may
say they are not the men you took them for.

Much Ado about Nothing, Act III, Scene 3

Science and technology make modern civilization possible. Without them, the living standards of developed nations could not be maintained nor the living standards in other nations improved. High standards of living depend increasingly on energy. In previous centuries, humanity's effect on climate was not of concern, if it was considered at all. Human influences on climate are now a critical issue. Suddenly abandoning fossil fuels is not feasible. As the chapter on fossil fuels implies, it can be assumed that before we are done, all the coal, oil, and natural gas humanity can get its collective hands on will be used up. The energy available from nuclear fission is greater than that from all fossil fuels put together and cannot be neglected, despite its hazards. But science and technology offer still more energy opportunities that have not been pushed to their limits. In the end it will take absolutely everything we can do to keep civilization going and increase the living standards of billions of people.

The chapters that follow describe energy possibilities that are achievable with current technology. Of course, fossil fuels will continue to be of great importance as long as they exist, and nuclear energy sources increasingly will be required as well. But as Chapter 10 shows, known technology allows energy to be used more efficiently than it is now. Enhancing the efficiency of coal-burning power plants is already underway and will become more important as time goes on. Increasing the efficiency of nuclear power plants is more difficult, but can be done.

In addition to fossil and nuclear fuels, there are many other, renewable ways to produce power, every one of which should be exploited to its limits. Part II presents an overview of these non-nuclear and non-fossil-fuel energy sources, either already in operation or otherwise possible with existing technology. But all have their problems as well as promises. Let us begin with solar energy, because without it none of us would exist.

Energy Demand and Climate Change: Issues and Resolutions. Franklin Hadley Cocks
© 2009 WILEY-VCH Verlag GmbH & Co. KGaA, Weinheim
ISBN: 978-3-527-32446-0

6
Solar Energy

Behold the blessed vision of the sun, no longer pouring forth his energies unrequited into space, but, by means of photo-electric cells and thermopiles, these powers gathered into electric store-houses to the total extinction of steam engines and the utter repression of smoke.

Mr. Punch, *The Telegraphic Journal and Electrical Review*, 1891

As Mr. Punch observed, most of the sun's energy ends up in the depths of inter-stellar space; the Earth captures only a tiny part of it. Even so, the energy the Earth receives from the sun each hour is more than humanity produces in a year from all fossil, nuclear, and renewable sources put together. As a renewable resource, sunshine is unparalleled. Despite the sun's tremendous output, the rate at which it generates energy per average cubic meter of its volume is actually less than the rate at which each human being produces heat. People usually generate from 50 to 100 watts per person from the food they eat, a light bulb's worth, and the sun generates less than one watt per cubic meter, averaged over its entire volume. But the sun encompasses a stupendous number of cubic meters. It is roughly 880,000 miles (1,430,000 kilometers) in diameter and occupies a volume more than one million times that of the Earth.

The sun produces energy by fusing light atoms together, primarily hydrogen, to make heavier atoms, especially helium, and loses mass in the process. Every second, the sun's nuclear fusion reactions convert about 4,000,000 tons of matter into energy! For comparison, a 1000-megawatt nuclear or coal-fired power plant takes one year to convert 0.130 kilograms (about four ounces) of matter into energy. Coal-fired plants are not normally thought of as converters of mass into energy, but in fact they are, as we saw in Chapter 5. The Earth is approximately 93 million miles (150 million kilometers) from the sun. At this distance, the sun-light shining on the top of the Earth's atmosphere delivers about 1366 watts per square meter (0.17 horsepower per square foot). The amount of this sunlight actu-ally received per square foot or per square meter at the Earth's surface varies with both cloud cover and latitude. Because the Earth is nearly spherical, the power in sunlight also varies with position on the Earth's surface, even if clouds didn't confuse matters. The sunlight falling upon each unit of area of the Earth's surface

Energy Demand and Climate Change: Issues and Resolutions. Franklin Hadley Cocks
© 2009 WILEY-VCH Verlag GmbH & Co. KGaA, Weinheim
ISBN: 978-3-527-32446-0

grows weaker as noontime shadows grow longer at ever-higher latitudes, north or south, from about 1000 watts per square meter at noontime in deserts near the equator to zero watts on the ice and snow during winter at the poles.

Because sunshine delivers more energy in one hour than humanity uses in a year, sunlight could certainly supply all human energy needs. Harnessing this enormous resource is the rub, because sunlight isn't available 24 hours a day on the Earth, except during summer near the poles, and is spread out unevenly over the planet. The problem is how to capture, use, and store the energy in sunlight. Plants, of course, have been doing this for a long time.

Sunlight is diffuse and spread out. Large areas are needed to collect a lot of it, whether for agriculture or for the direct conversion of sunlight into electricity using solar cells. The known technologies for harnessing the power in sunlight are manifold. Some are very simple, and some are not.

Using Solar Energy

Simple approaches to using solar energy include solar cookers and hot water heaters, as well as just basking in sunshine. Architects can design for solar heating and lighting or to minimize temperature gain in the summer. At some latitudes, awnings can be sized and positioned so that the slanting rays of the winter sun shine through windows, but summer sunshine does not. The photoelectric cells Mr. Punch mentioned, which are called solar cells or sometimes photovoltaic devices these days, are a step up on the technology ladder. The thermopiles he refers to are today's thermoelectric devices, and the bundles of thermocouples Mr. Punch called thermopiles are still sometimes used. Thermocouples are simply wires of two different metals, each pair of wires joined together at one end. When the connected end of a thermocouple is heated, a voltage appears across the unconnected ends. Thermoelectric devices operate as thermocouples do, but they are made from semiconductors, not metal wires. Figure 6.1 illustrates the physical difference between a solar cell and a thermoelectric device.

Because thermoelectric devices are driven by differences in temperature, their efficiency for converting unconcentrated sunlight into electricity is very low, much less than 1%. But concentrated sunlight focused on the top plate increases the temperature difference between the top and bottom of the bismuth telluride rectangular units and raises the conversion efficiency. If done properly with highly concentrated sunlight, the efficiency of thermoelectric devices for converting sunshine into electricity can approach that of silicon solar cells. Presently, most thermoelectric devices are used for cooling. When electricity is forced through solar cells, they just heat up. When direct electric current is forced through thermoelectric devices, a temperature difference is produced between its top and bottom surfaces. Both thermopiles and thermoelectric devices convert the heat from sunlight into electricity. Solar cells convert the energy in sunlight directly into electricity. Heat is a collateral product, not a driver, of the electrical power from solar cells. Heat is the driver of electrical power from thermoelectric devices.

(A)

(B)   (C)

Figure 6.1 (**A**) The physical appearance of a typical silicon solar cell. The top electrical contact wire connects to a central bus-bar solder strip that is itself connected to multiple side strips. This silicon cell is made from multicrystalline silicon, not single-crystal silicon, as can be seen from the light reflected from its different crystal grains. This cell has been treated with an antireflective coating to reduce the reflection of sunlight as much as possible. The electrical connection to the bottom is made by contact to the solder-coated lower surface of the cell. A cell such as this has an efficiency of about 10% for converting sunlight into electricity. Details of the interior structure of a silicon solar cell are shown in Figure 6.2. (**B**) The physical appearance of a thermoelectric device. The top and bottom surfaces of this thermoelectric device consist of thermally conducting ceramic plates. Sixteen rows and sixteen columns of rectangular bismuth telluride units are situated between the ceramic plates and are alternately connected in pairs at top and bottom. Alternate rectangular units consist of bismuth telluride with additional elements added to make their electrical conductivities either p-type (positive-charge carriers predominating) or n-type (negative-charge carriers predominating). Thermoelectric devices either produce direct electric current if one side is heated and the other cooled, or they become hot on one side and cold on the other if direct electric current is supplied to them. (**C**) A radiograph of the thermoelectric device shown in Figure 6.1B, showing the complicated interior structure of this device. In this radiograph, the alternate top and bottom connections of pairs of bismuth telluride columns cannot be distinguished because they overlap and appear as a continuous strip.

In the 19th century, thermopiles using the heat generated by concentrated sunlight could convert a greater percentage of the sun's energy into electricity than the solar cells of that time. Under optimum conditions, the best solar cells can now turn 30% or more of the energy in sunlight into electricity. There is hope that nanoscale thermoelectric devices will do better than bismuth telluride, one of the best present thermoelectric materials, but so far, solar cells are way ahead. Solar cells and thermoelectric devices generate direct current rather than alternating current. Of course, direct current can be converted into alternating current, but doing so adds its own cost. Unlike normal electric power generators, solar cells have no moving parts, and their useful lifetime is indefinitely long. They never need lubrication, for example, but they may need cleaning from time to time. Even the high-efficiency solar cells of the Mars rovers *Spirit* and *Opportunity* slowly lose effectiveness as they accumulate Martian dust, until a breeze from a passing dust devil periodically blows some of it away.

Widespread use of solar cells is inhibited by the investment required to produce each watt of electric power. On the Earth's surface, solar power is also handicapped because it is intermittent. A 1000-megawatt coal, gas, or nuclear power station needs only a few acres (or hectares) of land and operates day and night. A 1000-megawatt solar power station occupies square miles (or square kilometers) and works only in the daytime. Even so, the dream of producing electric power from sunlight on a massive scale won't die. Solar cells have been used to power satellites and space probes for decades. Someday, they may help to power a moon base. Unlike the situation on Earth, sunlight shines constantly in space, and the idea of deploying enormous solar cell arrays in near-Earth orbit and sending the electric power they generate to Earth using microwave transmitters was suggested more than 40 years ago by the Czech-American scientist and engineer Peter Glaser (see Chapter 21). One problem with all concepts for large-scale space solar power is their cost, which is extremely high. Combined with the technological dream of space elevators, truly large-scale generation of electric power in space for Earth and the "total extinction of steam engines" might become possible, but right now it is not.

Solar cells are economically favorable right now for generating electric power in remote locations, in other places transmission lines don't reach, or in situations where excess power can be sold to utility companies at appropriate prices. With the investment of a lot of money, the cost of solar cells should steadily decrease and their conversion efficiency increase, to the point where truly large-scale electric power production becomes practical right here on Earth. This process is already well underway. To get a feeling for what might be expected for the future of solar cells, it is worth looking first at their history and science.

Development of Solar Cells

In 1838 the French physicist Edmond Becquerel (father of Antoine Becquerel, who shared the 1903 Nobel Prize in physics with Marie Curie for the discovery of radioactivity) found that a voltage appeared between two pieces of platinum wire immersed in dilute sulfuric acid when one platinum wire was illuminated and the other was not. The power generated was minuscule, and the Becquerel photoelectric effect was only a scientific curiosity. Becquerel's electrochemical photoelectric cells are still not the best way for producing lots of electric power from sunlight.

It was not until W. G. Adams and R. E. Day published their discovery of solid-state solar cells based on selenium in the 1876–1877 *Proceedings of the Royal Society of London* that people like Mr. Punch began to dream of harnessing solar energy wholesale. Adams and Day described how a "little cylinder" of selenium with platinum wire contacts at each end produced electricity when only one end was illuminated.

In the time of Adams and Day, selenium was of research interest because of its use as electrical resistors in the circuits for transatlantic telegraph service. After great trials and tribulations, the first telegraph message had been sent across the Atlantic in 1858, but the failure of the insulation used for this first transatlantic cable soon ended service until another, insulation-improved cable was completed in 1865.

The transatlantic cable was one of the great engineering feats of the 19th century, and it stimulated all manner of research into improving long-distance telegraphy. Selenium played a part in this technological advance because at that time it was used in electrical resistors. But the resistance of selenium decreases when sunlight shines upon it. The fact that selenium was used in long-distance telegraphy and that its resistance depends on light intensity may have had something to do with Adams and Day carrying out the experiments they did. In any event, their paper soon led to the development of the first solid-state solar cells. These cells were made from iron sheets covered on one side with a thin layer of selenium that was itself coated with an extremely thin, partially transparent film of platinum. When the selenium layer is exposed to sunlight, electric power is produced and delivered via wires attached to the thin platinum film on top and the iron plate on the bottom.

These first selenium solid-state solar cells converted sunlight into electricity with an efficiency of about one-quarter of one percent, something like 1000 times better than Edmond Becquerel's original electrochemical cells. That's about where things stayed until the 1950s and the development of silicon solar cells with efficiencies of over 5%. Silicon solar cells were initially called solar batteries, even though they do not store energy. Quaintly, in the U.S. patent classification system, solar cells are still included in Patent Class 136 – Batteries: Thermoelectric and Photoelectric.

How Solar Cells Work

Solar cells accomplish the feat of turning sunlight into electricity by absorbing a photon of light near the contact region – the junction – either between a metal and a semiconductor or between different semiconductors or two types of the same semiconductor. The different behavior of the electrons in the two materials in electrical contact at the junction between them produces an internal electric field across this junction. This electric field arises because the number of electrons per unit volume and their ability to move around are different in each type of material. When these two materials are in electrical contact, electrons diffuse from one to the other, but they do so at different rates. This is what produces an internal gradient in the electric field across the junction region. This internal field separates the electrical charges that photons generate when absorbed near the junction. These separated charges can then recombine by flowing through an external electrical circuit, thereby producing electric power. Figure 6.2 shows the basic structure of a silicon solar cell.

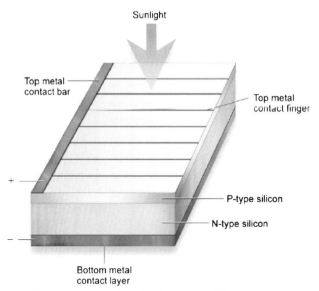

Figure 6.2 Schematic drawing of a classic silicon solar cell. Note that cells can have p-type silicon on top and n-type silicon underneath or vice versa. Different types of silicon are produced by adding very small amounts of certain elements. For example, adding boron to silicon makes it p-type (positive-charge carriers predominating). Adding arsenic makes it n-type (negative-charge carriers predominating). Solar cells can also be produced by using two entirely different semiconductor materials in contact, such as cadmium telluride with cadmium sulfide, or by eliminating the top semiconductor layer entirely and replacing it with a very thin, semitransparent metal or an electrically conducting oxide such as tin oxide. Finally, the semiconductors involved do not have to be crystalline and can consist of many large crystals instead, like the cell shown in Figure 6.1A.

Multiple-layer Solar Cells

By the 1950s space flight was just beginning, and the first satellite, *Sputnik I*, circled the Earth on 4 October 1957. Solar cells were an excellent way of producing power for satellites, and they still are. The era of satellites and silicon ushered in major technical advances that have boosted solar cell conversion efficiencies to well over 30%. However, reaching such levels of conversion efficiency requires fabricating complicated multilayer cells. These cells can cost about an order of magnitude more than regular silicon cells.

Solar cells work best when illuminated by a single optimum wavelength of light, but sunlight contains light of many different wavelengths. This is why a prism in sunshine produces multiple colors. There is more energy in a photon of sunlight with a short wavelength, such as blue light, than in one with a long wavelength, such as red light, and very short-wave ultraviolet light packs much more energy punch per photon than long-wave infrared light. The short-wavelength ultraviolet part of sunlight causes skin cancer, cataracts, and sunburns, because short-wavelength photons carry enough energy to break chemical bonds, which is why colored cloth bleaches in sunlight. The short wavelength part of sunlight also supplies most of the energy that plants use. Plants need the energetic short wavelength photons in sunlight to break the bonds in water and carbon dioxide so that they can reassemble these atoms into hydrocarbons. As wonderful as plants are, they make use of only about 3% of the energy in sunlight. It turns out that solar cells can do much better than this.

Extremely high-efficiency solar cells are made by stacking very thin semiconductor layers together, each layer aimed at a different portion of the solar spectrum, with the layers at the top selected to be activated by shorter wavelengths and those further down tuned to longer wavelengths. These layers must be arranged in this sequence because it turns out that short-wavelength cells will let long-wavelength light pass through, but long-wavelength cells won't let short wavelengths pass through. The resulting structure is complicated and hence expensive to manufacture. The different layers also need to be optimized for the specific parts of the solar spectrum they absorb. Even good, single-layer silicon solar cells have efficiencies usually less than 20%, although silicon cells can be improved a little beyond this if the cell is perfectly made. Of course, the rigid control of manufacturing procedures needed to make nearly perfect cells raises costs. Multiple-layer solar cells can reach over 30% conversion efficiency, but making them is very expensive.

The cost of solar cells alone, without counting the complications of packaging or the structure needed to hold them in place, is a major impediment to large-scale power generation. The main use of expensive, multilayer, high-efficiency solar cells is to power satellites, especially those in synchronous orbit, broadcasting television and radio signals to Earth. Located about 22,000 miles above the Earth, the signals from these satellites must be fairly strong, which means they need a lot of power, and they must be kept aligned as sunlight and the solar wind pushes them about. To keep their earthward orientation, these satellites use controlled magnetic fields to push against the Earth's field and generate the twisting-force torques needed to stay aimed toward Earth. Generating controlled magnetic fields takes power. If these satellites used some kind of small rockets to stay properly oriented, refueling from time to time would be needed. Magnetic orientation control by means of so-called magnetic torquer bars needs only the electricity that expensive multilayer solar cells generate so effectively. It costs a lot to launch anything into a geosynchronous orbit. For producing the same amount of power, extremely efficient multilayer solar cell arrays for geosynchronous satellites weigh a lot less than single-layer solar arrays would, making expensive multilayer cells economically viable for this application.

Multilayer cells use this same basic configuration of a basic solar cell except that several (usually three) cells are fabricated on top of one another. The electric current produced by the multilayer stack of cells is collected from the top of the first cell in the stack and the bottom of the last cell: The individual cell voltages then add automatically together, as the cells are connected in series by being stacked together, like the batteries in a long flashlight or pocket torch. Multilayer cells that are over 50% efficient have been produced experimentally. Typical commercial multilayer cells are over 30% efficient.

There are numerous technical factors that determine the efficiency of the sunlight-to-electricity conversion process, but the most important is this: If a photon of sunlight has more energy than is needed to produce a pair of electrical charge, then the extra energy is turned into heat instead of electricity. The more efficient a solar cell is, the larger the fraction of the energy in sunlight it converts into electricity and the less into heat.

The voltage that a single-layer solar cell generates is usually less than one volt. The power produced by a solar cell is equal to its voltage times the current it produces. To increase the low voltage, multiple cells are connected in series so that their individual voltages add up. Higher voltage is important because power increases as the voltage times the current, while resistive losses increase as the square of the current. By increasing the voltage, the same amount of power can be transmitted using a smaller current with a lower power loss. This is why giant power transmission lines operate at thousands of volts and hybrid gas-electric cars operate at hundreds of volts, not the 12 volts used by regular cars. What this all means for solar cells is that a lot of cells need to be connected together to make a solar panel that produces a reasonable voltage, such as 18 volts, instead of the one-half volt or so that each cell alone produces.

A solar panel must operate for a while just to give back the energy consumed in its manufacture. For silicon cells, at an absolute minimum, melting a kilogram of silicon starting from room temperature takes more than two million joules of energy, which is the amount of energy it takes to light a 100-watt light bulb for over 300 minutes. Suppose 10% efficient solar cells are produced that are about 0.01 inches (250 microns) thick. Generating the electrical energy it took to melt the silicon used to make such a cell, on a clear day in a desert near the equator, takes only a few days of operation. Actually, the energy payback is much longer than this, because it takes a lot of energy to convert sand into silicon, purify this silicon, grow a crystal, slice this crystal up (losing silicon in the process), and then process it to make solar cells and assemble them into panels. Estimates of solar cell energy payback time end up showing that a few years of solar cell operation are needed to generate the energy it took to produce them in the first place.

Thin-film solar cells, such as those made using amorphous silicon produced by the glow discharge decomposition of silane (SiH_4), the silicon analog of methane (CH_4), don't need as much energy to manufacture as crystalline silicon cells do and are only a few microns thick, but they are not as efficient as crystalline cells. Also, an amorphous silicon cell can degrade in sunlight, which is a very awkward characteristic for a solar cell to have. In spite of their problems, it is reasonable to expect that eventually large-scale generation of electricity will use some kind of thin-film cells, of which there are many types. In addition, thick-film cells several times thicker than thin-film cells have been made using electroplating methods, such as the electroplating technique developed for cadmium telluride–cadmium sulfide solar cells. Electroplating tends to be a cheaper production method than vacuum evaporation, which is the way that most thin-film cells are made. Thin-film cells can be cheaper in energy to produce than cells that have to be cut from crystals. From the big picture point of view, the lower the energy cost and the sooner the cell can return the energy used to produce it the better. Thin-film solar cells use a lot less semiconducting material than crystalline cells and might one day offer a renewable means of powering lighter-than-air ships and blimps in the place of internal combustion engines. It would be especially attractive from the engineering point of view if the entire outer surface of the airship could be made with thin-film solar cells. Of course, batteries or fuel cells would still be needed for overnight flights. The properties of batteries are considered in Chapter 11 and fuel cells in Chapter 16.

Many efforts are now underway to produce solar cells made using plastics rather than silicon. Other semiconductor solar cells besides silicon, such as gallium arsenide, cadmium-indium-diselenide, or copper-indium-gallium-sulfur selenide as well as cadmium sulfide–cadmium telluride are also being produced. Even solar cells made using combinations of inorganic and organic materials are being pursued. Most of these new cells are thin-film devices, because thin-film solar cells offer the possibility of costing less than crystalline cells, such as those made from crystalline or polycrystalline silicon. No matter how they are made or of what materials, the fundamental principles by which all these cells operate is essentially the same. An internal junction region between two materials with different electronic properties acts to separate the electrical charges produced when a photon of light is absorbed near the junction, and these separated charges can be made to recombine via an external circuit and produce mechanical work, electrolyze water, or do any of the many other things we use electricity for. It has also been proposed that nanoscale antennas could be used to turn sunlight into electricity, just as normal antennas are used to receive electromagnetic radiation. However, the super-tiny scale required for such antennas has so far made this approach impossible, except for very long-wavelength infrared radiation. If microfabrication techniques are developed sufficiently, however, the sunlight antenna idea might be more effective at converting sunlight into electricity than multilayer junction solar cells.

Most solar cell panels are flat. If they are not oriented perpendicular to the sunlight, their output decreases as the angle of the sunlight incident on them decreases, simply because the sunlight becomes more spread out at non-perpendicular orientations. It is certainly possible to continuously readjust solar panels so that they are always perpendicular to the sun's rays. Doing so, however, adds cost and complexity because of the steering and control apparatus required.

It is also possible to use reflecting surfaces to increase the intensity of the sunlight falling on a solar panel. If such reflectors are just flat mirrors, then these reflecting surfaces need to be steered so that they point toward the sun; otherwise, the shadow of the mirror itself will degrade power output as the sun moves across the sky. It is an optical fact that if a linear parabolic reflecting surface is cut in half along its apex and the two halves are separated and tilted equally away from each other, then any ray of sunlight striking one of the inner surfaces can be delivered, after multiple reflections, to a solar panel surface placed in the separated apex area.

Such complicated reflectors were developed for high-energy physics experiments involving the faint glow created when a particle moves through a transparent substance, like water, faster than light moves through that substance. This radiation is the optical equivalent of sonic booms that occur when a jet plane moves through air faster than sound does. This effect produces the blue glow that can be seen when looking down through water into the core of an operating nuclear reactor. If you ever get a chance to do this, be sure you take any pencils and pens out of your pocket before bending over to take a look! However, some light intensity is lost upon each reflection, and unless the sliced parabolic reflector is made extremely reflective and is precisely oriented, the gain in light intensity at the focus area is severely impaired. For high-energy physics experiments, the cost of making very accurate, highly reflective light collectors is not a major problem. For solar cells producing electric power, cost is a very important factor. So far, it appears that the most cost-effective means of harvesting electric power from sunlight using solar cells is simply to use flat panels fixed in the best possible sunward-facing direction.

It is also possible to make tubular silicon solar cells by using the edge-defined, film-fed growth (EFG) method developed about 40 years ago for producing shaped crystals of sapphire, including sapphire tubes as well as other shaped crystals. Tubular solar cells could be used in combined heating and electricity generation systems, but such combined heat–solar cell devices are not even close to being cost-effective, so the EFG method as applied to silicon is used primarily for producing flat silicon sheets for solar cells.

Plates of glass or plastic that contain a luminescent dye or pigment have been used to increase the output of solar cells. Increased cell output is achieved because the sunlight falling on the surface of the glass or plastic causes the dye or pigment to fluoresce. This fluorescent light is confined by internal reflection so that it is emitted at the edge of the glass or plastic, and this is where the solar cells are placed. Looking edgewise at such a flat plate of dye-impregnated plastic clearly shows the higher intensity of the luminescence around the edges of the sheet compared with its surface. It has been suggested that this technique could somehow be incorporated into windows and the solar cells placed in the frame that holds the glass, but, of course, the light coming through the window would then be tinted.

Luminescent sheets act to concentrate the energy of the sunlight falling on a surface. Converting sunlight into a single color by using fluorescent dyes can increase cell efficiency if the wavelength of the fluorescent light emitted by the dye is optimized. Of course there are problems. For one thing, organic dyes can degrade in sunlight. Also, there are various losses of light intensity in the process of getting it to the edge of the sheet. However, aiming may not be as important for a luminescent concentrator as it is for most other types of concentrators, but, even so, luminescent concentrators have not been employed extensively so far. Other types of concentrator systems have been developed, and large solar energy conversion systems based on concentrated sunlight have been studied for years, so let's look at them next.

Solar Concentrators and Solar Thermal Systems

Solar cells are not the only way of generating power from sunlight, and concentrated sunlight opens up additional possibilities. Focused sunlight can be used to boil water or cook food (or entertain mischievous children incinerating ants with a magnifying glass on a summer day). The heat from sunlight focused onto thermoelectric devices can produce electricity. Steam produced by focused sunlight can be put through a turbine to drive an electric generator, just as steam produced by burning coal can – no solar cells required. As early as 1913, a 50-horsepower engine in Meadi, Egypt, was operated with the steam from water passing through a pipe running along the focus of a linear parabolic reflector. Linear parabolic collectors look like long troughs with curved sides, and they focus sunlight into a line running along the focus on the inside of the trough. Concentrating sunlight works best if the skies are clear and if the collector is steered to track the sun. All the hardware for this costs money.

A solar technology that is cost-effective right now is solar water heating using unfocused sunlight. Just as the interior of a car becomes much hotter if the windows are closed while it sits in the sun, the water in a solar hot water system also heats up more if the black collector surface is provided with a glass cover. To increase temperature even further, two or more glass covers can be added, but because glass absorbs some of the light passing through it, adding still more plates soon does more harm than good. A sheet of normal window glass, when viewed on edge, can appear green or even black, because normal glass contains other things (especially iron) besides silicon and oxygen. Higher-purity glass is commercially available, and its lower absorption lets more sunlight through and improves performance, that is, makes the water hotter, but high-purity glass costs more than regular glass.

Temperatures can also be raised using selective absorber coatings. Such coatings have high absorptivity, meaning that they appear very black, but do not emit strongly in the infrared. As we have already seen in Chapter 3, increased carbon dioxide in the atmosphere decreases the rate at which the Earth emits infrared radiation. Surprisingly, selective absorber paints reduce the infrared heat radiation emitted by a hot water solar heater, not because they block infrared radiation but because they transmit it. The best selective absorber surface consists of a velvet-like layer of semiconducting tellurium microfibers deposited onto a gold surface. In visible light such a selective absorber appears to be extraordinarily black, but to long-wave infrared light the semiconducting tellurium is transparent, so that the emissivity is determined by the bright-and-shiny, low-emissivity gold surface beneath the tellurium. In this manner, surfaces that have absorptivity-to-emissivity ratios of over 30-to-1 have been produced. With ratios of absorptivity to emissivity this high, water in a solar hot water heating system can be brought to a boil without the need for focusing sunlight. The advantage of selective absorber surfaces is that they can be less expensive than solar concentrating systems, and they are not as dependent on orientation towards the sun. However, super-effective selective absorber surfaces are very expensive, and it is true that focused sunlight systems are needed when trying to reach temperatures a lot higher than the normal boiling point of water.

Solar Ponds

Without focusing, the thermal energy in sunlight can be collected and stored in ponds containing dense, very salty water on the bottom with fresh water on top. By making an artificial pond with a bottom layer of black plastic to increase the collection of solar heat, the salty lower layer just above the black plastic is warmed, while the top, fresher water stays cool, partly due to evaporation. The salty water does not rise up and mix with the fresh water because of its greater density even when warmed. The difference in temperature between the salty bottom water and the fresher surface water can then be used to operate a power-generating system based on chlorofluorocarbon or other fluids that boil at temperatures lower than does water. Such power plants have been built, but to date the cost of the electricity they produce is much higher than that from coal-fired power plants.

As the cost of fossil fuel increases, solar pond power plants deserve further study. They have a great advantage over solar cells because solar ponds stay warm after the sun has set and maintain a difference in temperature between the surface and the bottom water layers. But the efficiency of solar pond thermal power systems will always be much lower than that of coal or nuclear plants or solar cells, because the temperature differences they use are low (more about this efficiency difference in Chapter 9).

Solar-powered Air Conditioning

Using the heat from the sun to power air conditioning is obviously possible, simply by using the electric power from solar cells to run an air conditioner, but this approach makes woefully inefficient use of the energy in sunlight. First, most solar cells have only modest efficiencies. Second, most air conditioners use mechanical work to move heat from one place to another, the mechanical work being supplied by an electric motor. The coefficient of performance (COP) is defined as the ratio of the heat energy that is moved divided by the work energy needed to move it. For air conditioners in good working order and operating under normal conditions, the COP of an air conditioner is usually over 10. The so-called SEER value (Seasonal Energy Efficiency Rating) is just the COP adjusted for varying seasonal operating conditions, as the COP varies with outside temperature. If the size of the heat-exchange unit that discharges the heat from the air-conditioned building to the environment is increased, the COP and SEER values increase as well, but, of course, the cost also goes up.

Common air conditioners make use of the fact that when a gas is rapidly compressed it heats up, and when compressed gas is rapidly decompressed it gets cool. Some gases work better for this purpose than others, and nowadays most air-conditioning units use electric motors to drive compressors that operate with fluorocarbon or hydrofluorocarbon gases, but ammonia is also used in large systems. Compression is carried out at the point where the heat is to be discharged, and the resulting compressed gas, after it has been cooled off, is expanded at the point where the heat is to be taken up.

It is not necessary to use electricity to move heat around. After all, natural gas and propane-powered refrigerators have existed for a long time, and they use the heat from burning fuel to produce cooling. Refrigerators that burn fuel to operate are based on the cycle invented in the 1920s by two Swedish students, Carl Munters and Baltzar von Platen at the Royal Institute of Technology in Stockholm. The system they invented utilizes the fact that ammonia (NH_3) is very soluble in cold water but much less soluble in hot water, while hydrogen is hardly soluble in either and balances the pressure throughout the system. After ammonia is dissolved in cold water, a portion of this water is heated to produce ammonia under high pressure and elevated temperature. This high-pressure ammonia is cooled off by normal room temperatures outside the refrigerator. The resulting cooled high-pressure ammonia is then expanded through a nozzle in a tubular assembly inside the refrigerator, and the temperature of the ammonia drops as it expands rapidly from a high to a low pressure. It is this cold, low-pressure ammonia that then absorbs heat, keeping the inside of the refrigerator cold. The low-pressure ammonia then dissolves into water once more, and the cycle starts over again, the hydrogen keeping the system pressure balanced. This cycle operates steadily in moving heat from the inside to the outside of the refrigerator. The difficulty with ammonia–water–hydrogen is the need for temperatures like those produced by burning natural gas, and it's not easy to make solar heating systems that can produce high temperatures without using focused sunlight. Other heat-powered refrigeration methods, involving chemical combinations such as ammonia and sodium thiocyanate or water and lithium bromide, can work with lower temperatures than ammonia–water–hydrogen can, but, even so, they are not yet practical for solar-powered cooling due in part to their high cost and complexity. One great advantage of using solar power for cooling in air-conditioning systems is that the need for cooling is usually greatest when the sun shines the brightest.

Other ways of cooling using solar power are already in service. If the air is very dry, so-called air coolers, sometimes called desert or swamp coolers, work via the heat taken up when water evaporates. This cooling method is, of course, the one used by human beings when they sweat or by dogs when they pant. This method of cooling sometimes doesn't work very well, as anyone who has tried to mow a lawn on a hot, humid summer day knows. Blowing air around improves evaporative cooling, which is why fans are helpful if air conditioning isn't available. Air circulation in a dwelling can also be improved if a chimney is available, because hot air rises. Painting the part of a chimney that's in sunlight black helps this type of natural circulation. The chimney effect has also been proposed for power generation using the upwardly moving air to spin a turbine, as discussed next.

Solar Updraft Towers

There is another concept for generating electricity using solar energy that is completely different from solar cells; this concept involves a tower or, more precisely, a gigantic chimney. As everyone knows, a chimney channels the flow of heated exhaust gas and smoke from a fireplace upward and out of the house. It is also possible to use black collectors inside a greenhouse-like structure to heat air at the base of a very tall chimney and then funnel hot air up the chimney, because hot air rises. Interposing a wind turbine inside this chimney enables electricity to be generated from the heated air as it flows upward. One big problem is, naturally, the cost of the chimney. In 1982 a large prototype solar updraft tower was constructed and operated in Spain, south of Madrid. This chimney tower had a diameter of 33 feet (10 meters) and a height of 640 feet (195 meters). The collector area was 495,000 square feet (46,000 square meters). This device could produce a maximum of 50 kilowatts. After eight years of operation, it was shut down due to structural instabilities in the tower.

The ability of solar updraft towers to produce electric power at competitive costs has not been proven. One updraft concept that has not even been attempted is the generation of a tremendous updraft column of warm air using truly enormous collector areas together with a steep-sided mountain supporting a large, inclined tube-chimney of enormous length. Such a scheme was proposed in the 1920s by the French physicist Bernard Dubois and submitted to the French Academy of Sciences. Proposed for construction in North Africa, its "chimney" would have been more than a mile (1.62 km) long and 30 feet (9 m) in diameter. It was expected to produce updraft wind speeds of more than 100 miles (162 kilometers) per hour but was never built.

Natural updrafts produced without towers are common. Glider pilots, eagles, and condors frequently ride such updrafts to high altitudes. Nobody seems to have yet proposed a reasonable scheme for using natural updrafts to generate electricity. One big drawback of solar updraft towers is that when the sun sets, power soon ceases, unlike with solar ponds, whose power production can continue long after sundown.

Solar systems that use concentrated sunlight offer possibilities that aren't available with either solar ponds or updraft towers, so let's look at them next.

Solar Power Towers

Sunlight focused by reflection is the principle that Archimedes supposedly used to burn the Roman fleet at the Battle of Syracuse in 213 B.C,, as reported by one of the most renowned physicians of antiquity, Galen of Pergamon (129–circa 199 A.D.). Modern focusing collectors using reflected sunlight to produce intense light are sometimes called solar power towers, because the target for the reflected sunlight is usually mounted on a tower and the reflectors mounted on the ground. Solar power towers use assemblies of large mirrors directing sunlight to a receiver, where water is boiled at high temperature to drive turbines and spin electrical generators. Prototype systems having maximum capacities of over 10 megawatts have been built. Because high temperatures are used, the efficiencies of such systems can be comparable with normal power plants. The reflected light from each mirror must be kept targeted on the receiver in the tower. Problems of intermittent operation due to clouds and sunset are obstacles for such power systems.

It remains to be seen just how much electricity prices must rise before solar power towers become economically viable, but, even so, large prototype solar power tower systems have been built, particularly the one in the Mojave Desert near Barstow, California. First completed in 1981, it went through two modifications, called Solar I and Solar II, before achieving a power level of 10 MW with 1818 mirrors covering 780,000 square feet (72,500 square meters). Before being decommissioned, it had demonstrated that molten salts could be used to store heat energy for nighttime use. A third power tower, called Solar Tres, is currently under construction in Spain. This plant will also have a power level of about 10 MW, with sufficient molten salt heat storage to operate 24 hours per day. In all solar power schemes, cost is naturally a major factor. A standard solar power tower design would help lower costs. But each solar power tower project is currently unique. The solar power tower technique might become economically viable in favorable locales with lots of sun, but so far only prototype plants have been built.

Other Thoughts and Possibilities

In the last 50 years, the cost of electric power from solar cells has decreased by about a factor of 100. If the cost could be lowered still more, perhaps by a further factor of 10, it might become economical to produce electric power on a massive scale with solar cells, even though storing energy for use at night adds a lot to the final price. The use of solar energy for water heating is already practical and will become increasingly common as the cost of energy increases.

Some catalysts such as cerium oxide can produce hydrogen from water when exposed to sunlight, but the amount of hydrogen produced is trivial, just as the power produced by Becquerel's cells was. Some algae can do the same thing as cerium oxide and produce very small amounts of hydrogen when exposed to sunlight. Undoubtedly, research in this area could increase efficiency, but by how much is anybody's guess. Electrochemical methods for converting water into hydrogen and oxygen have long been known and are constantly being improved. If green plants could be genetically engineered to convert more sunlight into energy than they do now, biofuel production would be easier than it is currently (see Chapter 11). One especially interesting possibility is using genetically engineered, very high-growth-rate algae in conjunction with sewage treatment plants to simultaneously treat sewage, remove CO_2 from the air, and produce biomass that could be used to produce fuel. Another approach in using solar energy is artificial photosynthesis, which doesn't involve living things (see Chapter 18).

The development of stable, long-lasting thin-film solar cells should be pursued with all possible speed. Such cells should, of course, be as efficient as possible in converting sunlight into electricity and must be at least 7% efficient in doing this. Any solar cells that are less efficient than this are not very economically attractive for wide-scale power generation even if they cost nothing to make. The cost of land and the structures needed to hold these cells make low-efficiency cells unappealing.

Solar energy could become a mainstay of civilization, but to do so, the cost of harnessing it must be lowered substantially relative to that of fossil fuel. If 20% efficient solar cells could be produced as inexpensively as aluminum foil, solar cell use would explode across the world. There are few things as certain as the steady expansion in the use of solar energy. A resource that can supply hundreds of times more energy than humanity needs cannot be ignored, and it won't be.

Recommended Reading

Adams, W.G. and R.E. Day. "The Action of Light on Selenium." *Proc. Of the Royal Society of London* A25 (1877): 113. (Description of the first selenium solar cell.)

Backus, Charles E., ed. *Solar Cells.* New York: IEEE Press, 1976. (Classic coverage of early solar cell technology.)

Becquerel, E. "On Electron Effects Under the Influence of Solar Radiation." *Comptes Rendues.* 9 (1839): 561. (Becquerel's original paper.)

Cummerow, R.L. "Photovoltaic Effect in P-N Junctions." *Physics Review.* 95 (July, 1954): 16. (Description of the first silicon solar cell.)

Green, M.A. *Third Generation Photovoltaics: Advanced Solar Energy Conversion.* Berlin: Springer, 2005.

Nelson, Jenny. *The Physics of Solar Cells (Properties of Semiconductor Materials).* London: Imperial College Press, 2003.

Würfel, Peter. *Physics of Solar Cells: From Principles to New Concepts.* Weinheim: Wiley-VCH, 2005. (new edition to appear in February 2009).

7
Wind, Waves, and Tides

> *A favoring wind clear-eyed Athena sent,*
> *a brisk west wind that sang along the wine-dark sea.*

Homer, *The Odyssey*, circa 850 B.C.

> *Roll on, thou deep and dark blue ocean – roll!*

George Gordon, (Lord Byron), *Childe Harold's Pilgrimage,*
Canto the Fourth, 1818

This chapter presents the basic facts of wind, wave, and tidal energy, providing insight into what may be expected from these renewable energy resources as the world grows increasingly short of oil and troubled by what fossil fuel combustion is doing to the atmosphere. Wind, wave, and tidal energy will not solve all our energy problems, but they can certainly help a lot. The history of these renewable energy resources provides insight into their current status and potentialities. They are all valuable, but they all also have limits.

Long before the wine-dark sea of Homer's poetry, humanity knew the power of the wind, waves, and tides. Sailors harnessed wind power first, and only much later was it used to grind grain. The first written record of windmills was recorded by the Arabian writer Ali al-Tabari (834–927), in his story of a captured Persian slave/technician who claimed to be a builder of mills driven by the wind. Academic arguments over the role waterwheels played in windmill design may never be settled, but both windmills and waterwheels were widely used in the Middle Ages. Waterwheels operating with tidal water were a natural outgrowth of their river- and stream-powered cousins and were used for a long time, but generating significant amounts of electric power from waves is a fairly recent development. Let's look at all these renewable energy sources, beginning first with wind.

Wind

Only one-half of 1% of the energy in sunshine ends up powering the wind, but that's still a lot. Before the age of steam engines, wind alone propelled ships that plied the seven seas, and sailing ships still cross the oceans but no longer move much cargo. Commercial interest in wind energy has shifted decisively to the

Energy Demand and Climate Change: Issues and Resolutions. Franklin Hadley Cocks
© 2009 WILEY-VCH Verlag GmbH & Co. KGaA, Weinheim
ISBN: 978-3-527-32446-0

generation of electricity. As the price of oil increases, it is possible that energy for powering ships will eventually be supplied once again by coal, which even now has not been entirely abandoned for ship propulsion, or by nuclear power, long before wind-driven cargo ships become a common sight again. But auxiliary sails on engine-powered cargo ships have been shown to reduce total fuel consumption while not interfering unduly with delivery times. When the wind is not steady, the ship's engines can be used to drive it along, and when the wind is favorable, sail power substitutes for the engines. This application will very likely expand if the price of oil increases a lot.

The starting date for the beginning of sailing is lost in the mists of history, but the date of the world's first wind-powered electric generator is settled. The first wind-powered electric generator operated in Cleveland, Ohio, from 1888 to 1908, producing 12,000 watts of electricity used to charge the batteries powering 350 lights in the residence and laboratory of Charles F. Brush, an inventor and manufacturer of electrical equipment. Brush was a contemporary of Thomas Edison, and the Brush Electric Company ended up as part of the Edison General Electric Company in 1891, shortened first to General Electric and now to GE. GE is one of the world's leading manufacturers of wind power equipment and many other things, including the incandescent light bulbs that Thomas Edison invented in 1879 when he was 33 years old.

The first integration of wind power with electric utility service occurred in Russia in 1931 with the operation of the 100,000-watt Balaclava wind turbine utilizing the winds blowing in from the Black Sea. The world's first megawatt wind power system was the work of the American engineer Palmer C. Putnam, who talked the Morgan Smith Company of York, Pennsylvania, into funding the construction of a 1.25-megawatt wind turbine on top of Grandpa's Knob near Rutland, Vermont. It began feeding electric power to the Central Vermont Public Service Company on 10 October 1941. Although this wind power station operated for only four years, it held the record for the next 40 years as the largest ever built. It was a two-bladed, horizontal-axis device, not unlike modern wind machines, and the circle swept out by its blades was over 106 feet (50 meters) in diameter. The development of fatigue cracks, a bearing failure, and a shortage of parts during World War II made operation intermittent, until permanent failure occurred on 26 March 1945, when one of the blades fractured at its root as a result of fatigue cracking and was thrown far down the slopes of Grandpa's Knob.

Even though gearing and fatigue cracking problems have long since been resolved, additional problems remain. Most modern wind turbines are about 1-megawatt devices, but up to 50-megawatt units, over 50 stories high, are planned. Spinning blades can kill birds, but increasing the height of the towers used to support wind generators allows most birds to fly underneath their spinning blades. Wind turbines change the ambiance of the landscape or seascape. The variability of the wind itself and the cost of wind turbines versus the electric power they can produce are all important, and the engineering of wind turbines is highly developed. Wind is destined to play an increasing part in the world's energy supply. The engineering of wind turbines is utterly different than that for

solar cells, and it is interesting to consider some of the basic efficiency characteristics of windmills and wind turbines.

Characteristics and Limits of Wind Machines

The power in wind increases as the cube of its speed, because the energy in the
wind increases as the square of its velocity, and the amount of air passing through
the blades over any given time increases linearly with wind speed. As wind speed
doubles, the electric power it can generate increases by a factor of eight.

Theoretically, only about 60% of this power can be extracted no matter how
the windmill or wind turbine is designed, as Professor Betz at the University of
Göttingen proved in 1927. Betz showed that windmill efficiency is limited because
the wind must keep moving after it passes through the spinning blades. A windmill can only slow the air down; it can't bring it to a full stop, or no more could
pass through. The cubic dependence of power on wind speed makes it obvious
that brisk, steady winds are important.

There are many different designs for windmills. One basic point is that the
larger the width of the blades, the lower the wind speed that can start the windmill
going, but the larger the width of the blades, the more air friction they experience
when rotating, lowering the overall efficiency. In early mills used to grind grain
or pump water, large wind sails were used so that they would work even in gentle
breezes. In wind turbines for generating electric power, the blades are rather long
and narrow to increase efficiency, but this fact means that they don't start until
there is a decently stiff breeze. This is why modern wind turbines are frequently
seen to be stationary in mild breezes and why total efficiency is a strong function
of wind speed and constancy. If the speed is too slow, then the blades don't rotate,
and if it is too high, they must be feathered – turned edgewise towards the wind – to
prevent overstraining the generator, gears, and support tower. When the wind
blows steadily at the optimum speed, wind turbine efficiencies of 40% can be
reached, about two-thirds of Professor Betz's theoretical maximum. Additionally,
the wind blows anywhere it wants to, or it doesn't blow at all sometimes, and
as a result, the overall efficiencies of wind machines are about 20%, roughly half
of what they would be if the wind were brisk and steady. The evolution of the
engineering design of windmills over the last few centuries is illustrated by the
difference between very early Dutch and more recent Mediterranean windmills
contrasted with modern wind turbines, as shown in Figure 7.1.

In almost all commercial wind power systems, the shaft holding the blades is
horizontal to the ground, and the electric generator is mounted behind the blades.
The whole assembly can be made capable of rotating into the wind direction. A
large number of additional wind turbine designs have been proposed. Vertical-
axis, eggbeater-like wind machines have the advantage that the electric generator
can be on the ground, not on top of the tower. Vertical-axis machines don't need
to turn to face the wind. Of course, shafts and gears could be used to bring the
power produced by the spinning blades of horizontal-axis machines down to a

(A)

(B)

Figure 7.1 An ancient Dutch windmill as shown in a print by the famed Dutch artist Rembrandt van Rijn (1606–1669) (**A**). Existing windmills of similar design still exist on the Greek Island of Mykonos (**B**). In both cases, the windmill blades are intended to hold cloth sails, which are removed to stop the windmill from turning. These older designs are to be compared with that of modern windmills as seen from Copenhagen Harbor (**C**). In modern windmills, the blades can be rotated edgewise to the wind (feathered) to stop blade rotation.

(C)

Figure 7.1 *Continued*

ground-based generator. But in the end it seems to be better to eliminate as much gearing and shafting as possible and put the generator right behind the spinning shaft that holds the blades. Most commercial wind turbines are three-bladed, horizontal-axis devices on towers that spin electric generators placed right behind the hub of the blades.

At this point it is useful to step aside and say something about how electric generators work. They all operate on the principle discovered by the English physicist and physical chemist Michael Faraday. The son of a blacksmith, Faraday brought himself to the attention of Humphrey Davy, the discoverer of the elements sodium and potassium, by sending Davy a copy of the notes he had taken on one of Davy's public lectures on science. As a result he landed a job as Davy's assistant. Traveling across Europe with Davy, Faraday met many of the leading scientists of his day. In 1831 at age 40 he discovered that moving a coil of copper wire in a magnetic field produces an electric current. This is the principle upon which all electric generators rely to produce electric current. In power plants fueled by coal, nuclear, or natural gas, steam from boiling water spins turbines that rotate coils of wire in a magnetic field. In a wind machine, moving air spins the blades that rotate the wire coils. Because of the tremendous importance of generators, enormous engineering effort has gone into improving their efficiency, and generators can now turn into electricity over 90% of the mechanical power supplied to them.

To get an idea of the total supply of wind energy on the Earth, remember that each year sunlight supplies thousands of times more energy than humanity uses, and one-half of 1% ends up as wind that is potentially available for harvesting by wind machines. Potentially is the operative word, because a lot of this wind energy is almost as unavailable as is the enormous thermal energy locked within the Earth (see Chapter 9) or in temperature gradients in the oceans (see Chapter 15).

There is an enormous amount of power in jet-stream winds, which are bands of air moving from west to east at high speeds, typically around 100 miles (162 kilometers) per hour at altitudes of about 6 to 9 miles (10–15 kilometers). Remember that the power of wind increases as the cube of its velocity. Jet-stream winds occur primarily at latitudes between 30 and 60 degrees north or south of the equator. Jetliners frequently are helped by these winds when flying in the same direction and at the same altitude as the jet stream, which occurs commonly on major Northern Hemisphere airplane routes. Only a few percent of the power in jet-stream winds could supply all present electric power needs if it could be captured. But jet streams move around, in both position and elevation, and so far, harnessing them to produce power using kites or balloons has proven to be impossible, or at least not practical. Meanwhile, wind machines rooted in the ground have proven to work very well indeed.

At favorable locations where there are steady and strong winds, it can be profitable and practical to build and operate wind turbines to generate electricity. Low, rolling hills on large, flat plains are especially favorable because these hills tend to increase wind speed as air passes over them without turbulence. The American Great Plains is an especially favorable location for generating electricity using wind, with enormous potential that has yet to be exploited. The use of wind to generate electricity is growing rapidly and will have an increasingly important part to play in humanity's onward rush toward our energy destiny. Wind power is not going to supply all the electricity that people need, but it can safely be predicted to supply an increasing part of it. Other renewable energy resources are needed as well. We have already considered solar energy in Chapter 6.

Because more than 70% of our planet is covered by water, and wind blowing over water creates waves, some of the solar energy that ends up in wind is converted into the energy of waves. Up to now, the energy in waves has scarcely been used for electric power production. The development of tidal power is far behind that of wind machines but far ahead of wave power. Tidal power plants have already been built in favorable locations, so let's look at them next.

Tides

The early use of tidal waterwheels might have led directly to tidal plants for the generation of electricity if coal, oil, and natural gas did not exist. Tidal waterwheels were installed beneath the arches of London Bridge in 1580 and operated there for 250 years, before the advent of cheap power from oil and coal put them out of business. Global interest in tidal power has been reviving for more than 20 years as the benefits of this CO_2-free, renewable power source became evident. Where does the energy driving the tides come from?

Newton, the Moon, and the Tides

Chapter 2 discussed Isaac Newton's law of gravity and the fact that the Earth's equatorial bulge causes precession but did not mention the tides. In addition to spinning about its axis and rotating about the sun, the Earth also rotates around the center of gravity of the Earth–moon system. The Earth weighs about 100 times more the moon, and this center of gravity lies deep within the Earth, displaced about 3000 miles (4700 kilometers) from its center, in the direction of the moon. The moon can clearly cause one tide a day as its gravity pulls ocean water upward, but why two tides a day? Two tides occur because the Earth rotates about the center of gravity of the Earth–moon system, not the center of the Earth. A bucket filled with water can be swung around in a vertical circle without water falling out. The swinging-around motion of the Earth about the center of mass of the Earth–moon system acts similarly to displace ocean water on the side opposite to the moon, even as the moon's gravity acts to pull ocean water in the opposite direction.

Newton figured all this out and explained it in his book *Principia*, published in 1687. The sun has a similar but smaller effect, and both effects are somewhat merged together. Twice a month, at times of the new moon and the full moon, the tidal effect of the moon adds to that of the sun, and the resulting tides, sometimes called "spring tides," are high. They have nothing at all to do with spring, but that's what they are called, possibly because these large tides may seem to "spring" back and forth. Neap, or weak, tides occur when the moon is partially visible, neither full nor new, and the actions of the moon and sun tend to oppose each other. All this is confused by the time it takes for gravity forces to move ocean water about, by the fact that the Earth–moon and sun–moon distances are not constant, and by the fact that Jupiter has a small but detectable tidal effect. The whole thing gets very complicated but has nonetheless been thoroughly worked out, even though local tides are affected by the influence of the detailed geometry of the ocean–land boundary. In spite of these difficulties, accurate tide tables are available and provide the basic data for estimating the output that potential tidal power plant sites are capable of producing. The question then becomes determining how tidal power plants work.

Harnessing Tidal Power

Just like geothermal power, generating electricity using tides depends a lot on the natural circumstances that make certain locations more favorable than others. The configuration of the coastline is very important. It is especially helpful if a bay with a large difference in water level between high and low tide also narrows as it nears the open ocean. In this situation, the size of the dam-like structure, termed a barrage, needed to close the bay off from the sea is reduced. In the simplest type of tidal plant, power is generated when the ocean water flows through the generating system in the process of filling the bay as well as when this same water flows back out to sea at low tide. Naturally, in such a plant, the level of power generation

varies enormously each day as the tide ebbs and flows. If different bays are interconnected with separate barrages, power generation can be smoothed by alternately using the water flowing in and out of the different bays. A number of tidal
power plants already exist or will soon exist, most notably the 240-megawatt plant
at La Rance, France, and the 254-megawatt plant planned to begin operation in
2009 on the east coast of South Korea near Sihwa, about 40 kilometers south of
Incheon. The Shiwa tidal power plant will be the largest tidal power facility ever
constructed. During the Korean War, Incheon was the site of a major and successful amphibious landing by United Nations troops led by General Douglas
MacArthur, because the tidal range is about 30 feet (9 meters), and therefore this
coastline was not heavily defended. Two additional large tidal power plants, one
at Garolim (520 MW) and the other at Incheon itself (1440 MW), are also planned
in South Korea. Figure 7.2 shows the Shiwa tidal power plant as it will appear
when completed in 2009.

Much smaller pilot tidal power plants, of perhaps 10 megawatts or fewer, exist
in Annapolis, Maryland; Nova Scotia, Canada; Jiangxia, China; and Kislaya Guba,
Russia. In the 1930s a large tidal power plant had been planned at Passamaquoddy,
on the eastern seaboard of North America on the Bay of Fundy at the border
between Maine and Canada. Passamaquoddy is very close to Campobello Island,
where President Franklin Delano Roosevelt (1882–1945) owned a vacation home.
It was in this home that Roosevelt developed polio shortly after falling overboard
while sailing, becoming intensely chilled before being rescued. Roosevelt was very

Figure 7.2 The world's largest tidal power plant at Sihwa, on
the west coast of South Korea near Incheon, as it will appear
after construction is completed in 2009.

well aware of the gigantic tides in this region, which can easily exceed 33 feet (10 meters), and he was initially determined to build a tidal power plant there. The extremely large tides in the Bay of Fundy are caused by its enormous funnel shape. The very large Bay of Fundy grows narrower as its distance from the open ocean increases. Construction at Passamaquoddy had already been started in the 1930s when this project was cancelled, in part because of opposition to power generation under government control in New England as well as because of the argument that tidal electric power would be more expensive than that from the oil-fired plants that were widely used in New England at that time.

Usable Tidal Energy

The tides are gradually slowing down the Earth's rotation, about two seconds over 100,000 years. Knowing how much the Earth is slowing down, we can theoretically estimate the rate at which it is losing its kinetic energy of rotation. But not all of this energy ends up in the tides, because the Earth spins around every day, and the moon takes 28 days to go around the Earth. The tidal bulge is therefore dragged slightly ahead of the moon's vertical position with respect to the Earth's surface because of the rapid rotation of the Earth compared with that of the moon. The gravity of this tidal bulge slowly accelerates the moon in its orbit. As a result of this energy transfer from the Earth, the moon is moving away from us at a rate of about 1.6 inches (four centimeters) per year, which is about the same rate that the continents are drifting about on the surface of the Earth.

The Earth–moon distance could be determined accurately enough to measure the annual increase in this distance only after laser reflectors were left on the moon's surface as part of the *Apollo* lunar missions. It is possible to partition the energy loss by the Earth's slowing rotation into that delivered to the moon and that left in the tides. Going back to Newton and his law of gravitation, the energy effect of a four-centimeter annual increase in the Earth–moon distance can be calculated using the value of the gravitational constant first measured by the English chemist and physicist Henry Cavendish (1731–1810) at the beginning of the 19th century. A recluse, Cavendish was said to be the richest of philosophers and the most philosophical of the rich, having inherited immense fortunes from both sides of his family. His wealth enabled him to construct gigantic spheres of lead, along with railway lines to move them about, and a torsion balance of exquisite sensitivity. He used all of these things to determine the constant of proportionality for gravitational attraction (whose value was not known by Newton) in order to determine the mass of the Earth. Knowing the gravitational constant allows the direct calculation of the energy it takes to change the distance of the Earth to the moon, but it turns out that there are other, still more complicated factors as well.

Besides sorting out these gravitational effects, more confusing is the fact that the Earth's crust is slowly moving, the continents are slowly drifting, and the Earth's rotation is changing slightly due to the movements of its crust, just as ice skaters can spin around faster by moving their arms. The major ocean currents El Niño and La Niña in the Pacific Ocean also have small effects. As in so many natural phenomena, the whole matter is quite complex, but the final result is that the total tidal power on our planet amounts to about 5000 typical 1000-megawatt power plants worth, which is about twice the present electrical generating capacity of humanity. But almost all of the immense power in tides ends up as low-grade heat as the oceans slosh back and forth. The problem is to harness tidal power before it gets turned into low-grade heat. As in the case of hydropower, whose practicality depends a lot on geography, large tidal power plants need favored geographical locations, such as the South Korean coast near Incheon and the Bay of Fundy at the junction between the United States and Canada.

It is practical to harness the power in the tides only in those places where the configuration of the land and the sea is favorable. But even in locations where the tides can be used to generate electrical power, the power produced by individual plants is not enormous. Even the tidal power plants at La Rance in France and Shiwa in South Korea have only about 25% the output of a large coal-fired power plant. Tidal power, for all the theoretical potential it presents, cannot solve the world's energy problems because of the difficulty of harvesting it. But it can help. And it is certainly renewable.

Building and perfecting tidal power plants takes a lot of time and engineering effort, as well as money, but tidal power plants don't produce smoke, soot, or carbon dioxide – just electricity. About the only thing they alter is the way in which the tides go in and out. As energy supplies tighten, increasing attention should be paid to building tidal power stations in all favorable locations.

Tidal Currents

In addition to tides in combination with barrages to generate power, the high ocean currents that tides can produce in narrow straits and passages may also be used, without the necessity of constructing a barrage. When the potential of a tidal current power plant is evaluated, it is important to note that the energy available for harvesting increases as the cube of the tidal current speed, just as the energy in wind varies as the cube of the wind velocity. Up to a point, the faster the tidal current the better. For example, on the southern coast of South Korea, the Urdomag Strait, between Jindo Isi and the mainland, has tidal currents of around 12 miles per hour (five meters per second). Although it is extremely difficult to construct facilities in the midst of such rapidly flowing water, a pilot tidal current plant has been constructed there. The civil engineering difficulties encountered in constructing this tidal power plant in such strong currents can be appreciated from Figure 7.3.

Multiple tidal power facilities are planned for this strait. The energy of a wide channel of deep water moving at five meters per second is very large indeed. The engineering of facilities to harness some of this energy and turn it into electric power is challenging, to say the least, but the successful completion of the pilot tidal current plants at Urdomag Strait proves that doing so is not impossible.

Figure 7.3 The pilot tidal current power plant in the Urdomag Strait near the southern tip of Korea, showing the engineering difficulties in the construction of such a facility in the midst of rapidly flowing water.

However, the extremely high tidal currents in some straits might make construction of barriers to their flow too expensive to be practical. Each tidal strait must be evaluated individually to determine whether harnessing the energy contained in the rapidly moving water in the strait is practical. Numerous straits exist around the world, and it would seem valuable to begin serious consideration of harnessing this source of renewable energy. Small tidal current power plants have already been constructed at Lynmouth on the north coast of Devon in the United Kingdom (300 kilowatts) and in the Strangford Lough in Northern Ireland (1.2 megawatts). Just as for tidal power, power from tidal currents will not solve all of humanity's energy needs, but tidal power in all its forms is a significant renewable resource, and all suitable tidal power sites, whether involving barrages or straits, should be seriously considered as sources of renewable power.

In addition to wind and tides, the water waves that the wind produces as it blows across vast ocean distances can also be harnessed, so let's look at that possibility next.

Waves

Power production from wave energy is much less developed than that from wind or tides, even though wave power machines were patented in the 19th century. Like the energy in tides and wind, the energy of waves is spread out and not concentrated in one place. Harnessing it is not easy, but wave energy has long been successfully used for powering navigation buoys. More than 1000 wave-powered navigation buoys are in use. In buoy applications, wave motion can be utilized to compress air. The air compressed by the motion of the waves flows through a turbine to spin a generator to produce the power needed for signal lights, to produce an acoustic signal, or both. Acoustic or whistle buoys are especially useful in locations where fog is common. Compressed air produced by wave action turns out to be the approach taken in the world's first grid-connected, wave-powered electricity-generating plant in Norway, which became operational in 1985. This power plant is small and produces only about 500 kilowatts of power, about half of that generated by the grid-connected, wind-powered plant that used to exist on Grandpa's Knob in Vermont, but is easily accessible for maintenance and robust against storm damage. It is also provided with protruding harbor walls to direct incoming waves to increase plant output. In this aspect, it is similar to the tourist attraction called Thunder Hole in Acadia National Park on Mt. Desert Island in Maine, where a natural rock chamber lies at the head of a protruding rock channel that directs waves into a rock-walled air cavity, occasionally producing a booming sound as the water rushes in.

Another onshore wave-power concept is being developed on Kvitsøy Island, near Stavanger, Norway. This approach takes advantage of the fact that the height of a wave is greater than the average level of the ocean. Waves force water upward over sloping, horizontally slotted shoreline walls into several receiving chambers, from which it flows through multiple turbines back to the average ocean level. Multiple turbines reduce fluctuations in power as individual waves come and go. More than 10 megawatts of power per kilometer of coastline appear possible, and the concrete sloping walls are robust against storm damage. Perhaps such units could be built as part of coastal barriers, which are used anyway to protect harbors against storm damage. In any case, it would be wise to make the turbines and generators as waterproof as possible.

Onshore wave-power plants can be made much more resistant to storm damage than can those that float or are anchored to the sea bottom. In spite of their susceptibility to storm damage, dozens of ways to harness wave power using floating devices have been tried. Serious effort has also been made in Norway to develop large versions of wave-powered buoys to generate electricity on a large scale. Portugal is planning a 2.5-MW wave-power facility anchored offshore near the coastal town of Povoa de Varzim. This wave-power facility will have articulated sausage-like units, developed in the United Kingdom, and will use wave motion to pump high-pressure oil to drive electric generators. Offshore wave power offers the possibility of being a much larger source of power than onshore wave-power stations. But offshore wave power stations are vulnerable to storms, especially hurricanes, which can easily produce waves of such ferocity that these power-generation devices may be damaged, no matter how strongly they are anchored. It takes many small wave-powered systems to generate large amounts of electric power. But the energy density in waves is greater than that in wind because water is hundreds of times denser than air; therefore, even small wave machines can generate reasonably large amounts of power compared with wind turbines of the same size. The energy in a wave increases as the square of the wave height, so a steady supply of large waves is best.

If built on rocky shore, megawatt wave machines can be blended architecturally into the shoreline more easily than tall wind turbines can be melded into the ridgeline landscape. Naturally, there are more possibilities for power generation by anchored wave machines than from seashore installations, but floating power generators will always be threatened by ocean storms. Waves might supply a few percent of the world's electric energy needs. Wind power is already a big business, but wave power isn't. One handicap for onshore wave power plants is that coastal property tends to cost a lot more than land suitable for wind turbines.

Of the three sources of renewable energy—wind, waves, and tides—wind offers the greatest immediate possibilities, and it is not surprising that it is the most highly developed. Tidal power and tidal current power plants are also practical right now at favorable locations. Wind power is already being rapidly brought into service. Wave-power systems and tidal power plants have not yet been pursued worldwide to the same extent. But it may not be long before the need for renewable energy becomes so exigent that both wave power and tidal power will be pursued.

Recommended Reading

Berteaux, H.O. *Buoy Engineering*. New York: Wiley-Interscience, 1976.

Bernshtein, L. B., E. M. Wilson, and W. O. Song, eds. *Tidal Power Plants*, rev. ed. Seoul: Korean Ocean Research and Development Institute, 1999.

McCormick, Michael E. *Ocean Wave Energy Conversion*. New York: John Wiley, 1981.

Putnam, Palmer C. *Power from the Wind*. New York: Van Nostrand Reinhold, 1948.

Spera, David A. ed. *Wind Turbine Technology: Fundamental Concepts of Wind Turbine Engineering*. New York: ASME Press, 1994.

8
Going with the Flow: Water, Dams, and Hydropower

A man who has never looked on Niagara, has but a faint idea
of a cataract.
Mémoires de Bertrand Barère, 1843

Hydropower supplies more renewable energy than all other renewable energy sources combined and currently produces about 19% of all the electric power on Earth. It also has a long history, so let's begin with that.

Two thousand years before the Niagara cataract was harnessed to generate electricity, waterwheels were grinding grain. They are still being used to power various types of machinery, such as sawmills. Many industrial centers began where waterpower was available to operate textile mills, such as those that were built in Lowell, Massachusetts, in the 19th century to use the mechanical power that could be obtained from the Pawtucket Falls on the Merrimack River. Hydroelectric power was first used to light a dwelling in 1878, supplying the electricity for arc lamps (incandescent lights were introduced in 1879). Dams and a variety of modern versions of waterwheels, called water turbines, seem almost naturally adapted to the generation of electric power. Early dams and power stations were modest affairs, but they have grown ever larger as demand for electricity increases, especially from renewable resources. This growing demand compels the building of new hydroelectric facilities, both large and small.

One of the largest hydroelectric power stations on Earth is at Itaipú on the Paraná River, which divides Brazil and Paraguay. Completed in 1991 and upgraded in 2003, this hydroelectric facility supplies up to 14,000 megawatts (MW) of power. When completed, the Three Gorges Dam on the Yangtze River in China will be the largest source of hydroelectric power ever built, producing up to 22,500 MW, the renewable equivalent of more than 22 typical large, coal-fired power plants. Proposed by Sun Yat-sen in 1919, construction was begun in 1994 and is scheduled for completion in 2009. As large as these two hydropower facilities are, they represent only a fraction of the world's total hydroelectric power.

The rapid growth of electrical power from coal-fired stations is slowly reducing the proportion of Earth's electric power produced by water. Many of the best hydroelectric sites have already been exploited, but there still remain an enormous number of smaller hydroelectric possibilities, several major ones, and a few

Energy Demand and Climate Change: Issues and Resolutions. Franklin Hadley Cocks
© 2009 WILEY-VCH Verlag GmbH & Co. KGaA, Weinheim
ISBN: 978-3-527-32446-0

extremely large ones. The Tsangpo River, for example, is an undeveloped hydro-power resource in the Himalayas. It originates at Lake Manosarovar in Tibet and runs through Bangladesh (where it is called the Brahmaputra) before joining with the Ganges and flowing into the Bay of Bengal, running through enormous gorges en route. Damming this river could produce hydroelectric power on the same scale as the Three Gorges facility. There is no large, local demand for most of the electric power it could produce, and more than one country would be involved. But there is an enormous, pent-up demand for electric power in Asia. Twelve of the 17 largest hydroelectric plants under construction in the world are in China and will eventually produce nearly 90,000 MW of power.

Before being built, all hydroelectric projects, large or small, are analyzed for their potential. Besides environmental considerations, there are characteristics of water-power that are useful to know before trying to make an assessment of how well any hydroelectric project will perform.

Basics of Hydroelectric Power

When water, or anything else, moves from a higher to a lower elevation, its weight, moving through this height change, can be used to do work, such as spinning a turbine connected to an electric generator. At 100% conversion efficiency, it takes 100 kilograms of water falling a distance of about 20 feet (6.1 meters) to keep a 100-watt light bulb going for one minute. The higher the dam, the greater the power a given amount of water can produce. Only 50 kilograms of water would be needed to light that bulb for one minute if it fell through a distance of 40 feet. In some mountain lakes, water is sent through a pipe to a turbine at the base of the mountain so that the drop in elevation can be dramatically increased. Most small dams producing electric power are termed low-head hydro plants, where the term "head" refers to the height of the impounded water above the turbine. It is this height or head that determines the pressure with which the water flows through the turbines. For a given amount of water, the power that can be gener-ated increases proportionately as the head increases. The total energy that a given reservoir of water can produce with a turbine at its base increases as the square of the head, assuming a constant water surface area, because increasing the dam height also increases the total volume of water. In most cases, the surface area of the impounded water increases as the water level rises, so the size of the energy reservoir grows even faster than the square of the height of the dam.

The amount of total renewable energy that low-head hydro represents is diffi-cult to estimate accurately, but it might approximate that already developed from high-head, large-dam hydroelectric power schemes. One difficulty with low-head hydro is that the power produced is usually not sufficient to employ someone to supervise the operation, which must therefore be left on the waterpower equivalent of autopilot. Figure 8.1 compares and contrasts the dam facilities needed for low-head and high-head hydroelectric plants. High-head hydroelectric facilities are so large and cost so much that they require the participation of governments,

(A)

(B)

Figure 8.1 (A) An aerial photograph of Hoover Dam on the Colorado River at the border between Nevada and Arizona. When completed in 1936, it was the largest hydroelectric facility in the world. The hydrostatic head of water that drives its electrical generators is about 700 feet (213 meters). Hoover Dam produces about 2000 MW of power, the equivalent of two large, coal-fired power stations. (B) A low-head hydroelectric facility that can produce 2 MW of power from a 40-foot (12-meter) hydrostatic head. This low-head hydroelectric plant contains two generators, one within the dam itself and another fed from the overflow viaduct seen in the foreground.

whereas low-head hydroelectric facilities can be owned and operated by individuals.

Low-head hydropower plants entirely on private land may be more easily protected from damage or vandalism than those on public land. High-head power plants are always carefully protected. Even with their limitations, low-head hydropower plants can be expected to steadily increase in number, as the cost and demand for electrical energy increase and concerns about burning coal grow ever greater. Whether high-head or low-head, all hydroelectric facilities use turbines of one kind or another to convert the energy of flowing water into mechanical work driving generators to produce electric power, and these devices have their own special characteristics.

Water Turbines

Water turbines do not have the efficiency limit that wind turbines do. Wind turbines do not significantly change either the direction of the breezes flowing through or the pressure of the air. Water turbines usually change both the direction and the water pressure. Wind turbines are limited by a 60% theoretical efficiency limit because they can't significantly change the direction of the wind or the air pressure. Water turbines change water pressure greatly. Modern water turbines can be more than 90% efficient at turning the energy of flowing water into mechanical power. The original water turbines were simple waterwheels using either flat boards connected to an axle and pushed by water flowing underneath or wooden bucket catchments for water flowing from above the axle. These two types are called undershot or overshot waterwheels, respectively.

Modern water turbines are either reaction or impulse devices. In the impulse type, water squirts from a nozzle perpendicular to the axis of the turbine and is directed into bucket-like blades, pushing the turbine into high-speed revolution. In reaction turbines, water under pressure flows through piping, called penstocks, into the turbine itself, either parallel or perpendicular to the axis, driving impeller-like blades, which dramatically reduce its pressure. Unlike wind turbines, which are usually limited to a few megawatts in output, water turbines can be enormous. It is not unheard of for a single water turbine to generate over 100 megawatts of power.

Turbines can be designed to run either forward or backward, either generating electric power or using electric power to drive the turbine backwards as a pump. Such units are commonly used in conjunction with nuclear power plants, which run best when operated steadily. The excess electric power produced by a nuclear station running overnight can be used to pump water to a higher elevation and thus store the excess power for use when the demand for power is greatest (see Chapter 11). This principle can also be used in river hydropower schemes. In the Niagara Falls hydroelectric facility, for example, six turbines of the reversible type operate on the Canadian side of the falls, either to generate power or to pump water to a lake above the falls, where it can be used during times of peak power demand.

Hydropower Problems

All hydropower projects are susceptible to changing climatic conditions, especially drought. But remember, as global temperatures increase so too does the evaporation rate of water from the oceans. Within the coming century, as the average ocean temperature increases, the evaporation rate of water from the oceans might increase more than 20%, causing additional rainfall worldwide and thereby raising the potential for hydroelectric power generation. Remember that increasing the temperature from 77 to 78.8 °F (25 to 26 °C) increases the evaporation rate by about 6%. However, rain can increase in some places while at the same time decreasing in others, even while water evaporates faster from the oceans. Total worldwide rainfall will increase, but its distribution and type, such as drizzles versus deluges, also very likely will change.

Extremely large-scale hydroelectric facilities require years of planning and can have many effects on the local environment as well as on fish migration patterns, especially for salmon. Financial, political, and economic difficulties make large hydroelectric projects singular events for the regions in which they occur. Large projects can have extended, long-term environmental effects, as the Aswan High Dam across the River Nile has proven. Completed in 1970, it can generate more than 2000 MW of electric power, but even more importantly, it controls the annual flooding of the Nile. Thousands of people and many ancient Egyptian artifacts, including the Temple of Abu Simbel and its enormous statues of Ramses II, had to be relocated to escape the flooding caused by this dam, which produced the 300-mile (500 km) long Lake Nasser. Because the Aswan High Dam reduces the downstream annual deposition of silt, Egyptian agriculture has been affected, and the Nile Delta has been adversely impacted by increasing subsidence. Large dams are attractive targets in war or for terrorist attacks. The catastrophic failure of any of the world's major dams could easily take many more lives than did the disastrous failure of the Chernobyl nuclear reactor.

Hydropower Schemes

Perhaps the greatest potential source for hydroelectric power production on the planet involves damming the Congo River, forming a gigantic lake in central Africa and refilling Lake Chad to its size of 10,000 years ago. The hydroelectric potential of this scheme varies, depending on how it would be carried out, but it could be several times larger than the Three Gorges Dam. The Congo Lake project was first studied in the 1930s, when major parts of the African continent were still under colonial rule. The Congo River pours over 1.2 million cubic feet (34,000 cubic meters) of water *per second* into the Atlantic Ocean, flowing over a series of cataracts known collectively as Boyoma Falls in Zaire, formerly Stanley Falls. (It was Stanley who uttered the memorable greeting, "Dr. Livingston, I presume," upon finding that famous doctor in 1871 at Ujiji, in what is now Tanzania, on the shores of Lake Tanganyika).

In millennia past, the Congo River had formed a large lake filling the Congo Basin, until breaking through the surrounding hills to flow outward to the Atlantic. Damming the Congo River near its exit point from the Congo Basin could reform the Congo Lake to its original area of 350,000 square miles (919,000 square kilometers), submerging a portion of central Africa. Furthermore, diverting part of the outflow northward into the Ubangi River would refill Lake Chad, creating an even larger lake in central Africa. The scale of such a project is beyond colossal and is utterly unfeasible politically, but its hydroelectric power potential is staggering, as are the environmental and human problems building it would cause.

Many proposals have recently been put forward for Congo River hydroelectric projects considerably smaller than the plan proposed in the 1930s. There are already two moderate-power hydroelectric facilities in place. A third dam as planned would deliver 4500 MW of power. A fourth project, called Grand Inga, envisions producing almost 40,000 MW of hydroelectric power, nearly double that of the Three Gorges Dam. Connected to a gigantic power grid, it could supply a large part of the current electric power needs of the entire African continent. Grand Inga would be the largest hydroelectric facility in the world. Some have argued that it might be better, and far less environmentally disruptive, to supply electric power to rural African communities by means of a very large number of small, dispersed solar cell power stations rather than one colossal hydroelectric plant.

On a much smaller scale, the largest underground hydroelectric power plant on the North American continent is at Churchill Falls, Labrador, Canada. This plant has a capacity of 5000 MW, much of which is carried to the New England and New York power grids via high-voltage transmission cables through the province of Quebec. Although the Churchill River runs above ground, a combination of geology, terrain, and climate led to the decision to install the power station itself underground. Other notable North American hydroelectric dams include the Grand Coulee (6800 MW) on the Columbia River and the Hoover (2000 MW) and the Glen Canyon (1300 MW), both in the Colorado River system. The Nurek (3000 MW), Krasnoyarskaya (6000 MW), and Sayano-Shushenskaya (6400 MW) dams in Russia and the Guri (10,200 MW) in Venezuela are a few more examples of major hydropower facilities. The La Grand Complex in Quebec, Canada, generates over 16,000 MW. Worldwide, there are many other major hydroelectric plants, together providing almost one-fifth of the world's electric power. Massive hydropower facilities use dams, but it is not always necessary to build a dam to generate hydropower.

Dam-less Hydropower: Evaporation Schemes

In one form of dam-less hydropower, ocean water would flow through turbines into an area below sea level and then be removed by evaporation. There are several places on the Earth where the level of the land is below the level of the ocean. Death Valley, California, is one of them. At its lowest point it is 282 feet (86 meters) below sea level, but Death Valley is more than 150 miles (243 km) from the Pacific Ocean and is surrounded by mountains. In contrast, the Qattara Depression in Egypt is only about 40 miles (65 km) from the Mediterranean. At its lowest point, the ground level is 440 feet (134 meters) below sea level, and the total depressed area is about the size of Lake Ontario. Water from the Mediterranean could be brought most of the way to the Qattara Depression via open canal. In only one place does the land elevation increase enough that tunneling is necessary, through a distance of about one mile (1.6 km). The area of the Qattara Depression is about 7000 square miles (18,000 square kilometers). The air there is desert-dry and extremely hot, so the rate of evaporation once the depression is filled would be so great that it would take about 45,000 cubic feet (1300 cubic meters) of Mediterranean water per second to keep the newly formed "Sea of Qattara" from evaporating away. This is not quite half the average flow of the Niagara Falls that so impressed Bertrand Barère. (Barère was a French revolutionary who voted in 1793 for the execution of Louis XVI and was exiled for regicide in 1816 after the restoration of the French monarchy.)

Like the Dead Sea, on the border of Jordan and Israel, or the Great Salt Lake in Utah, the salt content in a newly created Sea of Qattara will increase over time. The inflow of water from the Mediterranean into the Qattara Depression would be great enough to generate steadily about 1000 megawatts of hydroelectric power, no dam required. This power level is equal to that from a typical large, coal-fired power plant. The Qattara Depression is entirely within the borders of Egypt, a fact that would reduce the political difficulties of its construction. And Cairo, with a growing demand for electric power, is less than 142 miles (230 km) away. The Dead Sea, whose surface is 1378 feet (420 meters) below sea level, offers another possibility for dam-less hydropower using evaporation, but the political problems of building such a power facility may be insurmountable.

No discussion of power production that makes use of differences in water level sustained by evaporation would be complete without mention of the so-called "Atlantropa" plan, proposed by the German engineer and architect Herman Sörgel in the 1920s. Sörgel also championed the full-scale Congo Lake plan. He pointed out that the total flow of river water into the Mediterranean, which is fed by major rivers such as the Nile from Egypt, the Rhone from France, the Ebro from Spain, and the Po from Italy, is not sufficient to maintain its water level. Without additional water, the level of the Mediterranean would drop about 5.5 feet (1.7 meters) per year. To maintain the level of the Mediterranean constant, more than 940 cubic miles (4000 cubic kilometers) of water are drawn in annually from the Atlantic Ocean through the Straits of Gibraltar. This flow rate is about 10 times the flow rate of Niagara Falls. Sörgel's scheme for building a dam between the Pillars of Hercules at the Atlantic entrance to the Mediterranean is colossal in magnitude, just as his Congo Lake plan was. Only in desperate circumstances is it conceivable that it would be countenanced, even though the resulting drop in water level would add many tens of thousands of square kilometers of new land, joining Europe and North Africa together, to form the continent Sörgel called Atlantropa, as well as producing thousands of megawatts of hydroelectric power. But there is another, less stupendous, way of generating electric power from the flow of Atlantic water into the Mediterranean, as we shall see.

Dam-less Hydropower: Flowing Water

In mountainous or hilly regions it is often possible to divert portions of large streams around rapids and bring the diverted water through a pipe down to a lower elevation. At the pipe outlet, this water drives a turbine and generator to produce electric power before being returned to the stream. The power produced is proportional to the fraction of water diverted from the stream. A process similar to this is used at Niagara Falls, the amount of diverted water being increased overnight but reduced in daytime to restore the falls for tourists. Where such water diversion is applicable, it is usually preferred to dam construction, which is expensive. These so-called run-of-river hydropower projects are commonly used because of their reduced cost compared with building dams. There is also another dam-less way to produce power from flowing water.

Just as wind turbines generate electric power from moving air, so too can water turbines generate electric power from flowing water. Water turbines can be made fully submersible and integrated with generators that do not extend above the water surface. A major test of this concept is planned for the East River in New York City, involving six small units, each rated at 36 kilowatts. Wind turbines can be of enormous size, but the depth of the water limits the size of submerged water turbines. One compensating factor is the fact that the density of water is about 850 times that of dry air at atmospheric pressure and room temperature, and the energy density of moving water is correspondingly greater than that of air moving at the same speed. The average velocity of water in the East River is only 4.5 miles per hour (7.3 km/h), about as fast as a man can walk, and the turbine blades will rotate at 36 rpm in this flow. This rate is slow enough that these moving blades are not expected to present a significant hazard to fish. It remains to be seen what corrosion and electrical problems develop because the generator units are under water. Presumably, such problems can be overcome by suitable engineering. What is not evident is the economics of power generated in this way.

The need to use multiple small, submerged units to generate power presents substantial maintenance problems. If the necessary engineering of waterproof generators is accomplished, and if this concept proves practical, a new approach to electric power generation opens up that could have broad, worldwide applicability. In particular, Herman Sörgel's hopelessly improbable plan to dam the entrance to the Mediterranean could be replaced by large underwater turbines making use of the natural flow of Atlantic Ocean water through the Straits of Gibraltar, and this could be accomplished without the gigantic political and environmental difficulties of his Atlantropa scheme.

Major ocean currents, such as the Gulf Stream, are also potentially gigantic sources of power, but currents in the open ocean can shift position, as the Gulf Stream routinely does. In addition, it is not clear how the power generated by submerged propeller/generator systems in the open ocean could be brought over long distances to shore. But shipboard, open-ocean power could be used in place to generate hydrogen by electrolysis, with the hydrogen then transported to the mainland, all of which would naturally add to the total cost.

On a smaller scale, there are many rivers on the face of the Earth, and many of them flow deeply and steadily. The total potential of the submerged water turbine approach to power generation is not readily estimated, but it is certainly large. Many possible hydropower possibilities have not yet been fully exploited. Additional waterpower possibilities are discussed in Chapter 15. The potential of water to supply energy is far from exhausted. But the construction rate of large coal-fired power plants dwarfs that of large hydropower plants, in part because coal-fired plants don't need special geological conditions. The supply of hydropower will steadily increase even as the fraction of the world's electric power it supplies steadily decreases. Nonetheless, as renewable energy becomes more important, hydropower will continue to play a major role as the world's seemingly insatiable demand for electric power grows with no limit in sight.

It would be a wise decision indeed for the world's governments to examine the remaining possibilities for hydroelectric power production, both large and small, that have not yet been exploited. Planning should begin now to examine systematically the remaining hydroelectric energy possibilities across the Earth. The exploitation of low-head hydropower should be encouraged by appropriate polices that make construction and operation of these modest power projects viable right now. A large number of additional low-head hydropower projects could add significantly to the world's supply of renewable electric power. The technology is already there. All that is required is appropriate legislation, which should be enacted without delay.

Recommended Reading

Ley, Willy. *Engineers' Dreams: Great Projects That Could Come True.* New York: Viking, 1954. This early book discusses a large number of engineering schemes relating to energy, giving information obtainable only with difficulty elsewhere.

Kaltschmidt, Martin, Wolfgang Streicher and Andreas Wiese, eds. *Renewable Energy: Technology, Economics and Environment.* Berlin: Springer, 2007.

Wilson, John R. and Griffin Burgh. *Energizing our Future: Rational Choices for the 21st Century.* Hoboken, NJ: Wiley-Interscience, 2008.

9
Geothermal Energy: Energy from the Earth Itself

The temperature I cried! Is it not a fact that it rises rapidly as one descends, and that the center of the earth is liquid heat?

Mr. Peerless Jones, *When the World Screamed*, 1929

The possibility of supplying all of humanity's energy needs lies right beneath our feet. The Earth is a repository of stupendous quantities of energy in the form of hot or molten rock. The difficulty is in making this energy resource available to generate electricity.

The supply of geothermal energy is somewhat analogous to the supply of uranium in rich ore deposits. Where nature has concentrated geothermal energy, it can be harvested profitably, just as uranium ore can be mined economically from rich ore deposits. Where natural processes have not concentrated them near the surface of the Earth's crust, uranium ore and geothermal heat are harder to obtain. Beyond the use of geothermal heat to generate electricity or to heat houses and buildings, one of the best possibilities for using the Earth as a geothermal resource is simply to use the ground as a reservoir for cooling as well as heating. We'll see later on in this chapter how this is possible. But first, let's look at using the Earth's heat directly. Where does this heat in the Earth come from?

Geothermal Energy

It took a primordial supernova to create the heavy elements in the sun and the rest of the solar system, as discussed in Chapter 5. Radioactive isotopes of uranium, thorium, and potassium produced by this supernova event are still present in trace amounts in common minerals such as granite and basalt, and the radioactive decay of these isotopes is responsible for most of the heat in the Earth. A small fraction of this heat also may result from compression of subsurface rocks due to the motion of the Earth's crust and earth tides caused by the sun and the moon, acting on different parts of the Earth's interior, which is not uniform.

Energy Demand and Climate Change: Issues and Resolutions. Franklin Hadley Cocks
© 2009 WILEY-VCH Verlag GmbH & Co. KGaA, Weinheim
ISBN: 978-3-527-32446-0

The Structure of the Earth

The inside of the Earth is not homogeneous, in either heat content or geology, and it is worth taking a look at its structure, which only began to be understood in the first half of the 20th century. The Croatian mineralogist and geophysicist Andrija Mohorovičić (1857–1936) was the precocious son of a shipyard carpenter. In addition to Croatian, by age 15 he spoke English, French, and Italian, later adding Latin, Greek, Czech, and German to the list. As director of the meteorological observatory at Zagreb, he developed an improved seismograph in 1908, just in time to observe the earthquake of 1909 in the Kulpa Valley, about 25 miles (40 kilometers) away. By comparing the measurements made with his and other seismographs located at different distances from the Kulpa Valley, he observed that some seismic waves arrived unexpectedly early. From this observation, he concluded correctly that the quake originated in the outer layer of the Earth, the crust, but that a deeper and denser layer, the mantle, transmits waves faster. Many seismographic observations have been made of numerous earthquakes since then, and the Earth's crust is found to vary in thickness, from about 4.3 miles (7 km) under portions of the oceans up to 22 miles (35 km) under the continents and 35 miles (60 km) under mountains. The boundary between the crust and the underlying mantle is called the Mohorovičić discontinuity. Figure 9.1 shows the overall structure of the Earth.

Beneath the crust lies the mantle, which consists of very dense rock. The mantle extends to about 1800 miles (2900 km) below the crust. The lower part of the mantle is a semisolid mixture of molten and solid rock. Underneath the lower part of the mantle lies the outer part of the core, which is mostly molten iron alloyed with some nickel and perhaps some lighter elements such as silicon. It is this molten iron–nickel layer that is primarily responsible for the Earth's magnetic

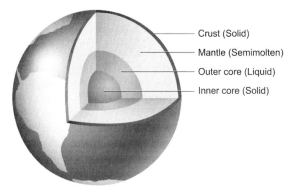

Figure 9.1 Schematic drawing of the structure of the Earth. The crust, upon which we all live, is extremely thin, ranging in thickness from 4.3 miles (7 km) under portions of the oceans up to 22 miles (35 km) under the continents and 35 miles (60 km) under mountains. In comparison, the distance from the surface to the center of the Earth is about 8000 miles (12,875 kilometers).

field. The core becomes solid again towards the Earth's center as a result of the tremendous pressure it is under, which compresses the inner core material enough to cause a liquid-to-solid phase change in spite of the high temperature, which is over 7200 °F (4000 °C)! We live on continents sitting on dense mantle rock that floats on a mixture of molten rocks floating on a molten iron–nickel alloy. Like most floating things, continents drift about. This motion happens not according to the human measure of time but to the Earth's, which is quite different.

It was the German meteorologist and scholar Alfred Wegener who first proposed, in the 1920s, that the continents move and change position on the surface of our planet. His studies led him to suggest that the Earth's landmasses had once been joined together in a super continent, which he called Pangaea, from the Greek for "entire earth." He reached this conclusion by noting that similar fossils exist on different continents and that some of the geological features of adjacent continents seem to fit together. Wegener's theory was met with considerable skepticism, and even ridicule. His death in 1930 during an expedition to Greenland to study polar air currents prevented him from living long enough to see his new idea accepted by the scientific community. Louis Agassiz, the discoverer of ice ages, had better luck. But Wegener was right; continents do drift, but only very slowly, about one inch (2.5 centimeters) per year, about the same speed at which the moon is receding from the Earth.

Molten iron–nickel and molten rocks, upon which the continents float and drift, hold unimaginably large quantities of energy in the form of heat. Only a tiny portion of this heat slowly seeps out through the Earth's surface, at an average rate of about 0.09 watts per square meter when averaged over the whole Earth, including the ocean floor. By comparison, the heat flow out of the interior of the moon is only about 0.019 watts per square meter, as determined by the instruments left implanted in the moon's surface by *Apollo 15* and *Apollo 17*. The Earth has many active volcanoes where molten magma seeps up through the crust. The moon no longer has active volcanoes that we know about, although a puff of hydrogen was recorded spectroscopically in 1958 issuing from the crater Aristarchus. If carbonate minerals are in the way of upward-rising magma, these minerals decompose and release carbon dioxide, which is a prime cause of terrestrial volcanic explosions. The molten interior of the Earth contains billions of times more energy than has ever been used by human beings, but no known technology can get at it, except in those places where hot spots extend upward, into, or through the crust. At such hot spots, geothermal energy is close enough to the surface to be usable, but not very many regions of the Earth's surface have enough geothermal energy to generate electricity in sizeable amounts. However, there are many places where small amounts of geothermal electric power can be produced. And geothermal power generation or geothermal heating for dwellings has long been used in those favored locations where the enormous heat energy resource in the Earth's interior approaches the surface and heats water.

Drilling for oil and gas is a highly developed technology, but no one has ever succeeded in drilling all the way through the crust, even though such a thing was proposed in the late 1950s and attempted in the 1960s. This effort, called Project

Mohole unsuccessfully tried to drill through the thin crust beneath the Pacific Rim off the west coast of Mexico. Project Mohole was funded by the U.S. National Science Foundation, which decided not to involve oil or gas companies in the project to avoid giving any of them a possible competitive advantage. But, of course, such companies are the most experienced in deep drilling. While Project Mohole did succeed in drilling under the ocean, what was mainly learned before it was cancelled was that it's very hard and expensive to drill to extreme depths, especially under water.

Drilling very deep holes even in dry ground is not easy or cheap. If it were cheap, the thermal gradient in the Earth is high enough to produce electric power simply by drilling down to depths where water would be near its atmospheric boiling point. Holes drilled several miles or kilometers almost anywhere on Earth will produce temperatures high enough to boil water at normal pressure, but there are two problems: drilling such deep holes is very expensive, and rocks don't conduct heat very well. Fracturing underground rocks using high-pressure water injection or down-hole explosions to increase the availability of hot rock to the injected water also costs money and energy. Without large-scale underground rock fracturing, the local rock is cooled by the injected water and would soon no longer heat it enough to generate steam. The main stumbling block is the cost and difficulty of drilling extremely deep holes into the crust of the Earth.

Virtually all the geothermally active portions of the Earth's surface are known, and two dozen nations currently produce electricity from geothermal steam. Some of the sites on the Earth's surface where geothermal steam is used to produce electric power are the Larderello steam field in Italy, which has an installed capacity of 500 megawatts (MW), the Wairakei field in New Zealand (160 MW), the Geysers field in California (2500 MW), the Tongonian field in the Philippines, and the Cerro Prieto field in Mexico (both about 700 MW) as well as smaller locations in Iceland, Japan, and several other countries. To put these operations in perspective, remember that a large, coal-fired power plant generates about 1000 MW.

The total world capacity for electric power production from geothermal steam would be large if cost were not a controlling factor. But even in some existing geothermal plants, capacity is degrading despite the continuous injection of water below the steam-gathering boreholes to try to maintain the steam flow rate. On the other hand, many new, smaller electric power–generating facilities in the one- or two-megawatt size range are coming online, and it is likely that the world's electricity-generating capacity from geothermal sources will at least double and possibly triple in a relatively short time. There are a large number of geological locations that can support small geothermal electric power facilities, and most of these have not yet been exploited.

With current technology, geothermal energy cannot supply a major fraction of the world's ever-growing need for electric power, but it can help. Injecting water into and below hot, dry rock can be used to try to create new fields, but this adds a lot to the cost. There are many more hot rock locations than geothermal steam sites, but how well they can perform if injected with water is not easy to estimate without expensive drilling and water injection experiments.

Geothermal energy is also not entirely free of environmental problems. Geothermal steam tends to be rich in hydrogen sulfide, the gas that gives rotten eggs their smell, and some electricity-generating geothermal steam systems have an unfortunate tendency to be somewhat odiferous, unless this sulfurous gas is re-injected into the Earth. Besides this environmental problem, geothermal steam has another, more basic problem – its temperature is not very high. The temperature of steam spewing forth from hot ground is low relative to that in coal, gas, or nuclear power plants. Typical power plants use steam at temperatures around 1000 °F (540 °C). Because geothermal steam has temperatures less than half of this, the efficiency at which the energy in this steam can be converted into electricity is also low, for unavoidable thermodynamic reasons deduced by Carnot at the beginning of the industrial revolution. Let's take a closer look at this critical factor in understanding the limits of turning heat into work.

Carnot's Unbreachable Thermodynamic Limit

The French engineer Sadi Carnot was the first to explain why the efficiency of steam engines increases as the temperature of the steam goes up. His life was foreshortened by cholera at the early age of 36. But his one and only publication, printed in 1824, has given him enduring scientific and engineering fame. Its title translates approximately as "Reflections on the Motive Power of Fire and the Machines for Developing this Power." This publication showed, correctly, that the maximum possible fraction of heat that can be converted into work cannot be more than $(1 - T_L/T_H)$. In this formula T_H is the high temperature of the steam entering the engine, and T_L is the low temperature (all in degrees Kelvin) at which the exhaust is discharged from the engine. More precisely, T_H and T_L are the high and low temperatures involved in the cycle through which the heat engine operates. The practical result of this thermodynamic limit to efficiency is that steam engines become more efficient as the steam temperature, T_H, is increased, thereby decreasing the ratio T_L/T_H. Geothermal steam is much lower in temperature than that in coal or nuclear power plants, which is why geothermal plants have low efficiencies.

Geothermal energy is economically practical for generating electric power only in those geologically favorable areas where molten rock from the Earth's interior approaches the surface. If an inexpensive way could be found to drill large holes to great depths, this situation would change, as the temperature at a depth of 20,000 feet (6010 meters) is usually around 400 °F (204 °C). But with current technology, drilling large-diameter holes to this depth is extremely costly and difficult. Even in favorable areas, only a small fraction of this heat energy can be converted into electricity because of its low temperature. Geothermal hot water for domestic heating is not affected by Carnot's efficiency limit, because for heating the energy in hot water doesn't have to be turned into work. Heating buildings, unlike generating electric power, doesn't involve heat engines.

Iceland is a volcanically active island with a small population and a lot of geo-thermal hot water, which supplies about 50% of its domestic and commercial heating needs during the long Icelandic winter, as well as heating its famous Blue Lagoon natural warm water bathing pool. There are many other places on Earth where water warmed by geothermal heat is available. Such natural sources of warm water have been valued throughout human history, perhaps none more so than the ancient Roman baths that still exist in Bath, England, a former spa for retired Roman legionaries who had completed their 20 years of service to the Empire. There are many more areas where hot or warm water is available from the ground than where steam is. The depth to which boreholes have to be drilled to produce warm water is much less than that needed to produce steam, and so the near-term potential for wide-scale geothermal heating is much greater than that for generating electric power from geothermal steam. But, electric power can also be generated using just geothermal hot water. This process has already been demonstrated in small (1-MW) facilities that use geothermal hot water and a heat exchanger to vaporize a fluid whose boiling point is much lower than that of water. Instead of water, these systems can use chlorofluorocarbons, hydro-chlorofluorocarbons, or ammonia, whose high vapor pressures are used to spin a turbine, after which the vapor is condensed again and recycled. Details of the warm-water-to-electric-power process are given in Chapter 15, which describes how warm and cold ocean water can be used to generate electricity. The Carnot efficiency limit for such systems will be low.

Drilling to depths of only 1000 feet (305 meters) or less is relatively inexpensive and can be used for air-conditioning or heating systems to improve their operation significantly. In-ground heat-exchange systems are also much quieter than air systems, which is important in densely populated areas. In-ground heat exchange represents a completely different approach to using geothermal resources, as we will see next.

Using Water and Soil in Heating and Cooling Systems

Water and soil can both be used to store heat. So far we've been talking about using the heat energy of the Earth to generate electricity, but there is another pos-sible way of using water and soil. Both can be used to decrease the energy cost of heating and cooling buildings. Let's first consider using soil or water in air-conditioning systems.

In most domestic air-conditioning systems, the heat being removed from the house is exhausted into the outdoor air, which is, naturally, at a higher temperature than the air inside the house. Pumping heat "uphill" takes more energy the higher and steeper the hill, that is, the outside temperature, is. The temperature of the ground a few meters down is approximately constant year round. If the heat being removed from the house is sent into cool ground, this process can be more effi-cient than shifting it into warm outside air. There is one problem: the ground does not conduct heat very well. Therefore, the coils and piping that put the heat into

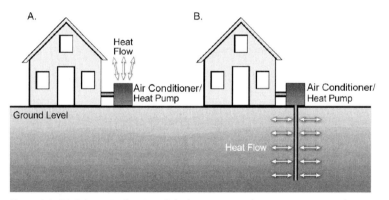

Figure 9.2 (A) Schematic drawing of the heat-exchange arrangement for a heat-pump system that heats or cools using air exchange. **(B)** Heat-exchange arrangement for a heat-pump system that uses the ground rather than air to exchange heat. In both systems, heat pumps, which can either discharge heat out or draw it in, are required. Using soil for heat exchange rather than air extends the range of climate conditions for which heat-pump systems can be used because of the constancy of the temperature of deep subsurface soil.

the ground have to be longer and more extensive than those that put heat into the air in order to spread the heat out and disperse it into soil. The feat of dispersing heat into the ground or drawing it from the ground is usually accomplished by drilling one or two holes about 1000 feet (305 meters) into the ground and installing metal or plastic pipes, which then transfer heat either into or from the soil by conduction. The use of such heat pumps, which are simply air-conditioning units that can be run backwards or forwards to pump heat into or out of a house, together with using the ground as a place to deposit or obtain this heat, will become more cost-effective as the price of energy increases. The operating cost of a ground heat-exchange air-conditioning system is lower than that for an air heat-exchange system, but the installation cost is higher. Figure 9.2 compares systems that exchange heat with the air (Figure 9.2A) or with the ground (Figure 9.2B) and shows their different structures.

Currently, it costs about three times more to install a ground-exchange heating and cooling system than it does to install an air-exchange system. A fan blowing across outdoor coils moves lots of air and is not very expensive compared with a large set of pipes installed deep in the ground. In spite of these factors, this technology is bound to be used increasingly, as such systems have much lower operating costs than the usual air-exchange systems. The overall operating cost of in-ground heat-exchange systems can be half that of heat-exchange systems that use air.

If an air-conditioning system is of the heat pump variety, then some of the heat exhausted into the ground in summer can be pumped back into the house in winter. Such systems are more efficient than heat pumps that use air, in part because the heat put into the air is lost. Geothermal heat-pump systems exist right now, but their high initial cost and the relatively low current price of electricity have inhibited their widespread application. Cooling systems that use water to increase the efficiency of air-conditioning systems are relatively inexpensive to install compared with in-ground systems, but they require nearby sources of cold or cool water. Some of these heat-pump systems use the cool water at the bottom of lakes for this purpose. By using water, not air, to carry away the heat, the amount of electric power needed to produce a given amount of cooling is substantially decreased.

The technology for all of these schemes exists today, but as in so many things, considerations of cost versus benefits dominate. Meanwhile, as the cost of conventional energy sources steadily rises, those who can afford the initial capital cost may opt to bury heat-exchange pipes underground to decrease heating and cooling costs. Greater exploitation of the enormous energy resource that lies beneath our feet requires greater governmental interest in this energy resource. Developing much cheaper ways of drilling to great depths should be a priority. For example, rocks are brittle solids, and therefore their fracture strengths can be reduced by adding the right things to their environment. Aluminum chloride as well as other chemicals, when added to drilling fluids, can substantially increase the drilling rate through hard rock, because these chemicals reduce the surface energy of rock. When rock is cut through, it is the surface energy that determines the ease with which cutting or fracturing can be done. Can other additives be found that will work even better? Even though much has already been done, there is a lot of experimentation still to be done in order to improve drilling technology, and the sooner advances are made the better. Better drilling technology would help in exploiting oil and gas fields as well as in advancing geothermal energy.

Recommended Reading

DiPippo, Ronald. *Geothermal Power Plants: Principles, Applications, Case Studies and Environmental Impact.* Amsterdam: Elsevier Butterworth Heineman, 2008.

Gupta, Harsh and Sukanta Roy. *Geothermal Energy: An Alternative Resource for the 21st Century.* Amsterdam: Elsevier, 2007.

10
Efficiency, Conservation, and Hybrid Cars

It is certain that the efficiency *can be made much more nearly to approximate* unity, *the limit of perfection, in electro-magnetic engines than in heat engines.*

William Rankine, *Manual on the Steam Engine and Other Prime Movers*, 1859

Using the energy sources we have more efficiently and reducing the amount of energy we currently waste are both destined to become increasingly important goals as the cost of energy increases. There are numerous ways in which efficiency can be improved. The low efficiencies of most electric power plants and internal combustic engines are especially important problems needing attention.

In 1859 the Scottish engineer William Rankine published the first complete analysis of the thermodynamics of power plants and predicted that steam engines can never be made as efficient as electric motors can. Rankine was right. Electric motors can turn more than 90% of electric energy into mechanical work. Power plants must convert heat into mechanical work and then use that work to generate the electric power that motors use so efficiently. Power plants are usually only around 35% efficient at turning the heat from the coal and nuclear fuels they use into electricity. Their efficiency is this low because they are heat engines turning heat into work, a process that has severe thermodynamic limitations, as we saw in Chapter 9. Internal combustion engines have the same problem. As a result, one of the greatest sources of energy waste is the energy thrown away as heat by electric power plants and internal combustion engines.

Improving living standards requires not only more energy but also more efficient use of the energy we have available. There are innumerable ways in which the efficiency of energy use can be improved. Many of these require an upfront investment, to be recovered later through lower operating costs.

The use of compact fluorescent lamps illustrates this situation. Compact fluorescent lamps cost more than incandescent lamps but use less electricity to produce the same amount of light. However, fluorescent lamps contain mercury, and incandescent lamps do not. There are about 50 milligrams of mercury in most standard fluorescent lamps for industrial or office lighting and somewhat less in smaller, compact fluorescent lights. Mercury is already a significant environmental

Energy Demand and Climate Change: Issues and Resolutions. Franklin Hadley Cocks
© 2009 WILEY-VCH Verlag GmbH & Co. KGaA, Weinheim
ISBN: 978-3-527-32446-0

problem, which the replacement of incandescent lamps by fluorescent lighting could worsen. Throwing fluorescent lights into the trash adds the mercury they contain to the environment, and mercury is hazardous to your health. Light-emitting diode lamps don't contain mercury, but they cost more than equivalent compact fluorescent lights. Energy-efficient devices tend to have higher upfront energy costs than inefficient ones.

More efficient use of energy is an added energy resource every bit as good as coal or oil deposits. What can be done across the entire spectrum of energy efficiency and conservation? Let's start first with those major sources of power that keep our civilization going – electrical power plants that use either fossil or nuclear fuel.

Efficiency of Fossil Fuel and Nuclear Power Plants

About two-thirds of the energy from the nuclear or fossil fuel used in power plants is currently wasted. Why is this so? One word reveals the answer: thermo-dynamics. In current power plants, water is heated and boiled at a constant pressure. In nuclear plants, the resulting steam is then expanded through a turbine without further heating. In most fossil fuel plants, after the steam is produced it is heated to a higher temperature before entering the turbine. This process is called superheating. Superheating allows fossil fuel plants to achieve about 40% efficiency, whereas nuclear power plants convert only about 32% of the heat produced by their nuclear reactors into electricity. Both coal-fired and nuclear power plants are limited by Carnot's efficiency limit for heat engines (see Chapter 9), and this efficiency increases as the steam temperature rises. Superheating is possible in coal- or gas-fired power plants, but it is very difficult to accomplish in nuclear plants. In nuclear power plants, water is used not only for steam but also to moderate the neutron flux that makes the reactor work. Superheating steam would be difficult and would involve changing reactors dramatically, at a stupendous cost.

Fossil fuel power plants can run with steam that is hotter than that in nuclear power plants, and that's why they can achieve a higher efficiency. In water-moderated nuclear plants there is no easy way to superheat the steam with nuclear energy because of 1) the rapidly increasing vapor pressure of water as temperature goes up and 2) the need to slow down (moderate) the speed of the fast neutrons from fissioning atoms. The need for liquid water limits the temperatures at which reactors can operate. Water is far safer as a moderator to slow down neutrons than is graphite, which would allow superheating. But water doesn't burn, and graphite does. The massive fire in 1986 at the nuclear power plant in Chernobyl, Ukraine, and the smaller fire in 1957 at the British Windscale reactor built to produce plutonium both illustrate the danger of using graphite as a neutron moderator. With graphite rather than water as the moderating material, the reactor can continue to operate while fuel elements are withdrawn in order to extract the weapons-grade plutonium they initially produce.

In both nuclear and fossil power plants, the steam exiting the turbine is condensed back to water at a pressure well below atmospheric. If cooling water from a lake, river, or the ocean can be used to condense low-pressure steam, condensation can be carried out at about 100 °F (38 °C). If a cooling tower is used, the condensation temperature is about 125 °F (52 °C). The efficiency is therefore a few percent greater in water-cooled plants. Also, it is a small but inescapable fact that as global warming proceeds, the efficiency of power plants will decrease slightly as the ambient temperature increases because of the slightly higher temperature at which the low-pressure steam can be condensed. The effect of outside temperature is clearly seen by the difference in power plant efficiency between summer and winter operation. The condenser temperature is lower in winter than in summer, so the overall plant efficiency is noticeably higher when the weather is cold.

The efficiency of fossil-fueled power plants can be raised by heating the steam by itself, after water has been boiled to produce this steam. This process is called superheating, and it increases the efficiency of plant operation. Basically, the steam is sent back into the firebox of the boiler using a separate set of pipes, as shown in Figure 10.1. This figure also shows the general arrangement of a typical fossil fuel power plant.

The caption for Figure 10.1 explains in detail the sequence of steps through which water goes as it is boiled, used to generate steam, condensed, and returned to the boiler in the Rankine cycle used in power plants to generate electricity. Chapter 9 has already explained why the low temperature of geothermal steam makes geothermal power plants a lot less efficient than coal-fired or nuclear power plants, which use higher-temperature steam than that available in geothermal power systems. All of these power plants are just giant heat engines and are inescapably bound by Carnot's efficiency limit for heat engines. Can anything at all be done to increase further power plant efficiency? The answer is yes, but it's neither easy nor cheap.

One way to increase power plant efficiency is to increase the pressure and the temperature of the water in the boiler until it is in the supercritical region. In the supercritical regime, there is no difference between the properties of liquid water and water vapor. In the supercritical region, water does not boil. The critical temperature and pressure points for supercritical water are 704 °F (373 °C) and 3200 pounds per square inch (psi) (22,000 kilopascals), respectively. But power plants can operate in the supercritical region at 1000 °F (538 °C) and pressures lower than 3200 psi because temperatures above 704 °F but less than 3200 psi can still remain in the supercritical region. The advantage of supercritical plants is the higher temperature and hence higher efficiency, usually around 40% instead of 35%. One disadvantage is that extraordinarily clean water is needed for supercritical plant operation. The boiler water has to be far cleaner in supercritical plants than in normal power plants because the accumulation of deposits from dissolved minerals in the boiler water cannot be cleaned from supercritical systems as easily as it can in normal power plants. Even so, power plants are sometimes converted from boiling water to supercritical operation to gain the extra 5% efficiency that supercritical power plants offer.

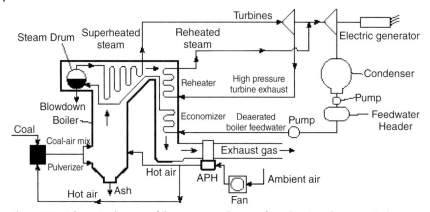

Figure 10.1 Schematic diagram of the steam, air, coal, and exhaust gas flows in a typical steam plant. The incoming coal is pulverized and mixed in the boiler with preheated air supplied by the air preheater (APH). The walls of the boiler consist entirely of water-filled pipes to extract as much heat as possible from the burning coal–air mixture. The hot combustion gas is then sent across superheater tubes to increase the temperature of the dry steam exiting the steam drum. This dry steam is generated in the steam drum from the wet steam produced by the water boiled in the water-filled tubes in the boiler walls. This hot, dry steam is then sent to a turbine and then through reheater tubes positioned in the hot combustion gas exhaust. After reheating, the steam is then sent through a second turbine, both turbines driving the electric generator, and then passes to the condenser, where it is cooled back to water using heat exchange with river, lake, or ocean water or with air in a cooling tower. The water exiting the condenser is deaerated, pumped up in pressure, mixed with any needed makeup water, and sent to the economizer for preheating by the hot exhaust combustion gas before being pumped back into the boiler again. Despite all these steps, most fossil fuel plants are around 35% efficient, which means that about 65% of the heat from the burning coal is wasted and does not generate electricity.

Another way to increase plant efficiency is to add gas turbines driven by burning natural gas or liquid fuel to the fossil fuel plant, to produce what is called a topping cycle. Topping cycle turbines are similar to jet engines, and their combustion gas temperatures far exceed 2000 °F (1100 °C). The topping cycle turbine drives a generator and produces electricity, and its exhaust gas is directed into the coal-burning boiler firebox to help boil the water and superheat steam. In this combination, the topping cycle turbine provides power, and the exhaust gas is used to replace heat that would otherwise have to be generated by burning more coal. Such topping cycle plants combined with superheating can achieve about 50% total efficiency, but they are not yet very widely used, in part because they are more expensive and complex than normal power plants and because the fuel for gas turbines costs more than coal.

One possibility for increasing efficiency is burning coal with water and a limited amount of air. This is the process that is used to produce so-called producer gas, which is a combustible mixture of carbon monoxide and hydrogen. For years such gas was used for cooking and heating in most industrialized countries. In New England, before natural gas pipelines from Louisiana finally reached that far north, producer gas was widely used. In a plant designed to produce both producer gas and electric power, the heat from burning wet coal under limited-air conditions boils water, while at the same time the producer gas generated supplies the fuel needed for the topping cycle gas turbines. The result is 50% efficient power plants, not the current 35%. Topping cycle power plants that generate their own producer gas could have a major beneficial impact in decreasing national and international carbon footprints, but developing them and putting them into widespread use will cost dearly. The way things are now, they won't be developed, because the power they produce will cost more than that from 35% efficient plants using coal. A carbon tax might change this situation.

Another possibility exists: power plants with efficiencies over 50% can operate without topping cycle turbines if a fluid with a lower vapor pressure than water, such as mercury, is used. The higher the vapor pressure, the greater must be the thickness of the walls of the pipes that contain this pressure and the poorer the rate at which they can transfer heat. For a plant that uses water, high temperatures mean that thick-walled boiler tubes are needed to contain the water vapor pressure. Thicker tube walls don't transfer heat to the water inside the tubes as effectively as thinner wall tubes do. Power plants that boil mercury rather than water could use boiler tubes with thinner walls as well as operate at higher temperatures, thereby yielding greater efficiency. But a power plant using tons of mercury presents such a dramatically serious environmental hazard that boiling mercury power plants are not used. Mercury presents a severe environmental hazard if it somehow escapes, and some of it always does. Massive amounts of mercury were used in the Oak Ridge Atomic Energy Commission facility, because it was needed to prepare lithium-6, an isotope that interacts with neutrons to produce tritium, which hydrogen bombs must have. Lots of this mercury is still there, dispersed in the water and soil around Oak Ridge.

Still another future possibility for a topping cycle is the combination of a high-temperature fuel cell that reacts coal with air without combustion, generating electricity in the process. As with the exhaust gas from topping cycle turbines, the exhaust gas from high-temperature fuel cells could then be used to boil water, replacing the need for some of the coal used to boil the water in the Rankine cycle portion of the power plant. Because fuel cells are not heat engines, and therefore are not inhibited by the Carnot limit, it is possible that power plant efficiencies well over 50% might be achieved. Fuel cells and their potential applications are discussed in more detail in Chapter 16.

Finally, it is worth noting that reducing pollution from fossil fuel power plants, which is a good thing, also seems to reduce plant operational efficiency slightly, thereby increasing coal consumption and carbon dioxide production, which are not good things.

At current coal prices, it is cheaper to burn more coal and produce electricity in the usual way and forget about trying to achieve 50% or more efficiency. It is also cheaper still to forget about pollution control, and some countries take the approach that the lowest-cost electricity is the highest priority, never mind pollution control or global warming. Some developed countries have thought it better to pay more for electricity and cut down on acid rain, soot, and other nasty things such as selenium dioxide and mercury vapor that flow out of coal-fired power plant smoke stacks. Soot settles out of the air rapidly, but carbon dioxide spreads around the entire globe. In almost all countries, no charge is imposed for letting carbon dioxide escape. That's the point. This situation is reminiscent of the old problem of common-pasture grazing of livestock. Individual farmers can benefit by grazing as many sheep and cattle as possible on the common land, but eventually overgrazing kills all the grass for everybody when it is eaten down to the ground. Similarly, individual nations can benefit by ignoring the worldwide level of carbon dioxide in the atmosphere. But unlike a denuded common pasture, whose grass can grow back in a few years, reducing atmospheric carbon dioxide to pre-industrial levels will take centuries, even if no extra carbon dioxide were added to the air.

There is still another way to make use of the heat that power plants waste, and that is to use this heat to warm homes, office buildings, and factories. This approach is usually called cogeneration. A cogeneration power plant's waste heat is used for space heating rather than being dispersed into air or water. Such plants can make use of 80% or more of the energy produced from burning coal, some for generating electricity and the rest for space-heating purposes. District heating is a similar concept involving a central supply of hot water for heating, but without any associated electric power generation. To be practical, district heating requires a high-density population living near the power or heating plant, such as on college campuses or in industrial complexes; otherwise, the loss of heat and the cost of pumping water over large distances become too great. Installing cogeneration or district heating systems is costly, but the energy savings are large. District heating, fueled by coal, has long been used on a large scale in Europe and Russia as well as in China, in cities north of the Yangtze River.

Cars, Trucks, Trains, Ships, and Planes

Except for sailboats, bicycles, skiing, or just plain walking, virtually all modern means of transport rely on the burning of fossil fuels. The economics of airplane operation has already forced reductions in fuel consumption per passenger-mile. Much smaller improvements have been made in the fuel requirements of trains and large ships. Most important of all, however, are cars, simply because there are so many of them. Until recently, the low cost of gasoline (petrol) in many parts of the world provided little or no incentive to improve the number of miles a car can go on one gallon or one liter of gasoline. We are already in the process of seeing what happens as oil becomes more costly. Cars with better mileage offer an economic advantage in fuel costs. If enough fuel-efficient cars cannot be put into service, then speed limits can be adjusted downward, as they were during the oil crisis of the 1970s, when speed limits were reduced to 55 mph (89 km/h), because air resistance increases as the square of velocity and lowering speed a little decreases air friction a lot. Reducing speed limits can reduce fuel consumption significantly. But in practical terms there are limits to how far the speed limit can be lowered.

It is obvious that small cars get better gas mileage than large ones and travel further on the same amount of gasoline. But when a small car and a big car crash together, or a truck and any car collide, other things being equal, the people in the larger vehicle will more likely survive. Safety features such as seat belts and air bags help with this issue. The larger the vehicle, the shorter the distance it will be able to travel on any given amount of fuel – that's the basic rule. However, there is something else that can be done: use hybrid power sources to propel vehicles. These hybrid vehicles can be demonstrably more efficient in their use of fuel at any size, especially in city driving, than their non-hybrid equivalents. In the 1970s oil crisis, hybrid vehicles were widely studied, and many schemes were advanced concerning the nature of the power source used to supplement an internal combustion engine. The three that emerged as most interesting were compressed air, flywheels, and batteries. All of these will work, with different limitations and problems, but it has turned out that batteries are the most practical choice by far, at least up to now.

Hybrid vehicles with batteries to supplement the power available from a small gasoline engine are at last commercially available. The increasing use of hybrids is influenced heavily by the cost of gasoline, but there are few things more surely written in the book of heaven than that the cost of gasoline will increase over time, up-and-down fluctuations notwithstanding. Hybrid vehicles are here to stay, but they work best for stop-and-go city driving, not long-distance, high-speed highway travel. Beyond hybrids, something radical will be needed to change the dependence of vehicles on gasoline. Chapter 16 explores what can and cannot be done with hydrogen and fuel cells.

Increasing the fuel efficiency of planes, trains, and ships in a way comparable to the use of hybrid cars is not practical with current technology. Planes can fly further if their weight is reduced and their already-efficient engines and wings are improved, but the possible efficiency advances still left to be made appear modest. Aeronautical engineers have worked for decades to improve airplane performance, and nautical engineers have done the same thing for ships.

Trains face a different efficiency problem. It is well known that trains can carry more cargo further using less fuel when the track is level and has fewer grades. But flattening out railway lines is extremely costly compared with the fuel savings that can be produced. This was demonstrated in the former Soviet Union, where the cost of interest was set at zero, meaning that enormous upfront expenses could be justified. Huge sums were spent on making train tracks as level as possible, no matter what the cost, because the resulting operating expense was thereby reduced. But this reasoning neglects the fact that time and money used for one purpose cannot be used for another. Time and money spent on making railroad tracks extremely level could be spent instead on building more efficient factories and homes.

The mileage of trains could be enhanced by substituting lighter materials such as aluminum for steel, even as aluminum is being replaced by lighter plastic materials or magnesium–silicon alloys in portions of automobiles and airplanes. A second approach to oil shortages and increasing gasoline prices is carpooling. A third is public transport and urban planning. Some cities have sufficiently extensive public transport systems that a private car is not necessary. Effective, efficient, and economical public transport usually requires a high population density. People living outside the city where they work, and the resulting suburban dispersal of population, make good public transport systems less economically viable. As the cost of private transport increases, urban living patterns can slowly adjust. Given sufficient time, the reality of high private transport costs can be accommodated by more extensive use of public transport. Rapidly increasing gasoline costs cannot be dealt with smoothly and may present a danger to the public order. The sooner the long-term inevitability of rising gasoline cost is recognized and planned for the better. These approaches to energy efficiency and conservation are very important, and whole books have been written about them, but they are not technological problems. There are, however, many technological approaches to energy conservation, so let's consider them next.

Conservation

Just like increased efficiency, conservation can be considered an energy resource. Conservation doesn't have to mean doing without. It is true that giving up things like cars and air conditioners would dramatically lower energy costs, but conservation can also mean just making better use of energy resources. An example is recycling rather than discarding aluminum cans and plastic bottles. "Waste not, want not" and that old Yankee axiom "Use it up, wear it out, make it do, or do without" still ring true. Naturally, recycling takes more effort than just throwing things away. But recycling can fend off shortages, as well as simplify the problem of waste disposal. It takes a lot more electricity to produce a fresh pound or kilogram of aluminum than it does to produce the same amount by recycling. To paraphrase Benjamin Franklin, a watt saved is a watt generated. As with urban planning, recycling tends to be more of a sociological problem than a technical one, but technology has a major role to play in the conservation of energy use.

Conservation is almost synonymous with efficiency. One example is the use of additional insulation in refrigerators. Some countries have mandated improved refrigerator insulation as a way of reducing electrical energy usage, but, of course, there are billions of people who don't have refrigerators at all. Energy standards for air-conditioning systems have also been introduced, but again, most of the world's people don't have air conditioners. The same is true for the efficiency of furnaces, which can be as high as 90% (meaning that 90% of the heat produced by fossil fuel combustion, whether natural gas or oil, is delivered into the home), but the great majority of existing furnaces deliver 70% or less for heating, with the rest going up the chimney. As the billions of people who do not presently have refrigerators, furnaces, or air conditioners begin to acquire such things, the world's demand for energy will increase, and efficiency of energy use will become even more important that it is now.

Speculation-built houses tend to have less insulation than custom-built houses because adding insulation adds to the builder's cost, while heating and cooling bills are paid by the buyer, who may not know what they are getting for their money until it is too late. Once again, there are a lot of people who live in houses with no insulation at all. It turns out that in cold or hot climates, where heating or cooling is necessary, the single most rewarding energy investment is more insulation. It is possible and practical in new construction to dramatically lower heating and cooling bills by using more insulation. In homes that are already built it is still possible, though more difficult, to increase insulation. Adding more insulation above attic ceilings is easier than installing it inside walls, and the savings from reduced gas and electric bills can be significant.

Projecting future natural gas and electric rates decades ahead is almost impossible, except for the truism that they will increase. There are intangible factors in conservation that are important even if they don't involve energy. For example, extra insulation also keeps out street noise, and using less natural gas and electricity decreases global warming. However, the economic value of such things to the typical homeowner is nowhere near as easily evaluated as the monetary factors involved in adding extra insulation or installing double- or triple-pane windows.

For homes and factories, the goal of energy audits is a reduction in energy usage. Sometimes energy can be reduced at little or no cost, for example, by turning off lights or agreeing to let the power company install a remote cutoff on air-conditioning units. In other cases a significant initial investment must be made to achieve increased efficiency of energy use, and a cost–benefit analysis is needed to determine what to do. Money used for energy conservation results in either lost interest or loan repayments, which must then be balanced against the cost of the energy saved. Some projects, while they save energy, may not seem to be worth what they cost until the overall loss in future energy savings is factored in. Efficiency, conservation, and hybrid vehicles all have increasing roles to play as time goes by and the cost of energy rises. The process of increasing efficiency and conserving energy has already begun. Hybrid cars are now available. It would be very helpful in controlling climate change and lowering energy demand if everything possible is done to increase energy efficiency. This is a critical matter that should be pursued at every level.

Recommended Reading

Boschert, Sherry. *Plug-in Hybrids: The Cars that will Recharge America.* Gabriola Island, B.C.: New Society, 2006.

Brown, Lester R. *Plan B 3.0: Mobilizing to Save Civilization.* New York: W.W. Norton, 2008.

Hofman, Konrad A., ed. *Energy Efficiency, Recovery and Storage.* New York: Nova Science, 2007.

Rankine, William. *A Manual on the Steam Engine and Other Prime Movers.* London and Glasgow: University of Glasgow, 1859.

11
Energy Storage: Macro to Micro

*On the Electricity excited by the mere Contact of conducting
Substances of different kinds.*

Letter from Alessandro Volta read to the Royal Society, 26
June 1800

Storing energy is not the same thing as producing it. But the ability to store energy, especially in large quantities, can help us to make efficient use of the energy we have. Let's look at how energy can be stored, on both the macro- and the micro-scales. There are many ways of storing energy, ranging in magnitude from colossal to microscopic. Almost all energy-storage methods use gravity, electromagnetic, or nuclear forces. Because gravity is the weakest of these, hydropower energy storage, which relies on gravity, requires large volumes of water falling through substantial distances to store large amounts of energy. Electromagnetic forces, while more powerful than gravity, are much weaker than nuclear forces. Consequently, coal and oil, which are in some sense repositories of stored chemical energy, must be far more massive than uranium or plutonium as a means for storing the same amount of energy.

Releasing energy by burning coal with oxygen produces heat, of course, but one difficulty is turning this heat into useful work, a process that is fundamentally bound by the Carnot efficiency limit discussed in Chapter 9. Stored nuclear energy has the same problem. To turn the energy stored in uranium into electricity means using a heat engine. Breeder reactors (discussed in Chapter 13) make use of nuclear forces to store enormous amounts of energy by producing more nuclear fuel than they consume. In this process, energy is stored in the form of the newly created nuclear fuel plutonium. Nuclear forces are millions of times stronger than electromagnetic forces and are unimaginably stronger than the force of gravity. Knowing this fact makes it understandable why breeding a few kilograms of fissionable plutonium stores more energy than the largest pumped-water energy-storage system on Earth. On the other hand, hydroelectric energy-storage systems are not limited by Carnot efficiency considerations when they convert their stored energy back into electrical energy. Currently, coal, oil, or nuclear power-generating plants are typically about 35% efficient, but pumped-water hydroelectric systems are usually 80% efficient. Let's see how pumped hydropower systems work.

Energy Demand and Climate Change: Issues and Resolutions. Franklin Hadley Cocks
© 2009 WILEY-VCH Verlag GmbH & Co. KGaA, Weinheim
ISBN: 978-3-527-32446-0

Pumped Hydropower

Pumping water from a lower to a higher elevation stores the energy produced by electric power–generating stations operating overnight when demand is low. This energy is then available the next morning by letting the water flow back down through turbines when demand increases, as people wake up and begin the day. To have an idea of the magnitude of the energy that can be stored by pumped-water systems, consider the 1080-MW hydroelectric facility in Northfield, Massachusetts. When it was constructed in 1972, it was the largest hydropower storage operation in the United States. This facility uses water from the Connecticut River, pumping it up 800 feet (244 meters) to a reservoir on top of a nearby hill.

This reservoir can hold 5.6 billion gallons (21.2 billion liters) of water, an amount great enough to provide 10 hours of full-power service. The power to pump the water into the reservoir is supplied by the surrounding grid-connected power stations that have excess capacity available overnight. The process of alternately storing and releasing energy from pumped-water energy systems is widely used to provide load-leveling for electric power–generating plants. Load-leveling means that the demands of peak periods of power usage, such as early morning and early evening, can be met by storing the excess energy available at other times. Energy-storage systems using water have been around for a long time, and a large number of pumped hydroelectric energy-storage systems are in use worldwide. All pumped-water energy-storage systems naturally require lots of water plus a nearby hill or mountain with a large lake or reservoir on top of it. Such things are not always available. It would be hard to make pumped-water energy storage practical in a flat desert.

Besides water, compressed air also can be used to store energy. Of course, this method doesn't need a lot of water, and air is available everywhere on Earth, but compressed air systems have other special requirements, so let's now discuss compressed air energy storage.

Compressed Air

The energy that can be stored using compressed air can be on a scale similar to pumped-water energy systems. The amount of energy stored depends on the volume of air used and its pressure. The amount of stored energy in a compressed air system increases as the pressure goes up, just as the energy in a hydropower system rises with the height of the pumped water. High-pressures require very strong containment systems to hold the compressed air, and for very large systems there is no practical alternative to underground storage. The walls of any compressed air tank must be thick enough to contain high pressures, and for a given storage pressure, the thickness of the tank walls increases as the storage volume of the tank increases. Big tanks containing high pressures need very thick walls. Storing large volumes of pressurized air in caverns deep underground solves the wall thickness problem.

Compressed air has an advantage that pumped-water systems don't. This advantage is the fact that compressed air can be used in conjunction with combustion turbines. A significant portion of the power of combustion turbines is taken up by the energy required to produce the compressed air that these turbines need in order to operate. If the compressed air required by a turbine is supplied from a source of stored compressed air, instead of from the compressor section in the turbine, the fuel needed to operate the turbine is substantially reduced.

This principal comes into play in the compressed-air energy-storage system in McIntosh, Alabama, built in 1991. In this facility, compressed air from an underground storage chamber is mixed with natural gas to supply a 100-megawatt gas turbine power-generation facility. This compressed air is contained in a salt cavity 2500 feet (762 meters) underground with a volume of 10 million cubic feet (283,200 cubic meters). The maximum pressure of the air in this chamber is 1100 pounds per square inch (7600 kilopascals). The power needed to compress this air is supplied by off-peak power from normal power plants. With compressed air, the fuel consumption required to run power-generating combustion gas turbines is 25% lower than for those without it. However, the overall efficiency of compressed-air energy-storage systems is not as high as the efficiency of pumped-water energy-storage systems. Why is this so? Remember that water is almost incompressible whereas air is very compressible. As air is rapidly compressed, it heats up. When pumping water to a higher elevation, it hardly warms up at all. As air heated by the process of compression is fed into an underground storage facility, it eventually cools down, losing some of the energy used to compress it. As a result the overall efficiency of compressed-air energy storage is lower than that of pumped-water energy-storage systems. On the other hand, if lakes, rivers, or reservoirs at different elevations are not at hand, then pumped-water storage isn't practical. If an underground cavern rather than a mountain is nearby, a compressed-air energy-storage system can be used to preserve some of the excess energy produced by power stations during times of reduced power demand. The amounts of energy that can be stored using compressed air are limited only by the size and depth of the underground chamber needed to contain the compressed air. If the geology is favorable, such chambers can be very large, like the one in McIntosh.

On a far smaller scale, the use of compressed air for powering cars has also been evaluated extensively, and compressed-air cars may be close to becoming commercial. This application is dogged, however, by the wall-thickness problem of the required compressed-air storage tanks. The greater the pressure, the thicker the walls must be to contain this pressure, and the larger the tank, the greater its wall thickness must be even if the pressure isn't increased. Using many small air tanks instead of one large one is handicapped by the fact that a larger total wall area is needed to store a given amount of compressed air in small tanks than in one large one. The more tank wall material there is, the heavier the total system, even if the walls are not thicker, leading to the problem of somehow optimizing the size and number of tanks. In spite of these difficulties and the danger that an exploding compressed-air tank represents, cars driven by compressed air have been built and tested. It is claimed that a very small car powered by compressed air might be able to go up to 125 miles (200 kilometers) before its air tanks need refilling. Such limited-range vehicles are suitable for city driving. From the global energy perspective, however, they suffer from the fact that compressing the air for storage in the tank inevitably leads to heating this air up and wasting some energy in the process, just as it does in underground compressed-air systems.

As an example of compressed-air energy storage, consider a 42-gallon (159-liter) compressed-air tank. Consider further that this tank holds air compressed to 5000 psi (34,474 kilopascals), which is very high as compressed air tanks go. The energy obtained when this tank is decompressed to atmospheric pressure is only about 1% of the energy that a 42-gallon gasoline tank can hold, assuming that the air expands adiabatically when it is used, meaning that it expands fast enough to not absorb heat from the environment as it expands. If the expansion of the air were a lot slower, then the work that can be obtained from it increases due to the heat it absorbs from the environment during the expansion process. Slower, however, means lower horsepower and worse vehicle performance. Rapid expansion of compressed air is closer to the true situation for the operation of a compressed-air motor than is slow expansion.

On the other hand, gasoline is burned in a heat engine (a car's internal combustion engine), and in the end only about 25% of its energy ends up in propelling the car. For an automobile that gets 30 miles per gallon (12.75 kilometers per liter), this means that the gasoline in a 42-gallon tank can propel the car about 1260 miles (2040 kilometers). If air-powered engines gave only this same 25% efficiency, then that 5000-psi, 42-gallon compressed-air tank could power the car for only 12.5 miles (20.25 kilometers). But compressed-air engines can have much higher efficiencies than internal combustion engines and can be over 75% efficient. At 75% efficiency, the compressed-air car could go 37.5 miles (60 kilometers). To reach the 125-mile (200-kilometer) figure claimed for compressed-air cars means they must have very large compressed air tanks and weigh as little as possible. Don't forget that compressing air takes energy, and in any national energy picture, this energy cost is important and needs to be taken into account.

It remains to be seen whether or not cars powered by compressed air are truly practical and able to compete with battery-powered cars. We'll look at batteries next.

Batteries

Batteries are not practical for storing energy on the same scale as hydroelectric or underground compressed-air storage schemes, but they can store enough energy to power cars, at least for modest distances. All batteries convert chemical energy into electricity without combustion and are therefore not limited by Carnot efficiency considerations (see Chapter 9). The Italian physicist Alessandro Volta (1745–1827) was a contemporary and friend of the Italian physiologist and anatomist Luigi Galvani (1737–1798). It was Galvani who had observed that the muscle in a freshly dissected frog's leg would twitch when in simultaneous contact with iron and copper, which form a crude battery. Galvani's observations led to decades of fruitless speculation on the nature of the "vital force" of life, but they also stimulated Volta into investigating the behavior of combinations of metals in various solutions. In 1800 Volta published the first description of a true battery in the *Philosophical Transactions of the Royal Society*. Volta's batteries involved disks of silver and zinc separated by pieces of leather soaked in salt water. When many such silver disk–wet, salty leather–zinc disk assemblies are stacked together with the dry sides of the silver disks in contact with the dry sides of zinc disks, the resulting "Voltaic pile" produces a voltage high enough to be detected without the use of any instruments. A Voltaic pile can deliver quite a shock. Unfortunately, Volta's batteries are not rechargeable. In this respect they are like common flashlight (pocket torch) batteries, which use zinc and graphite, not zinc and silver, and are also not rechargeable to any significant extent.

Only with the discovery of lead–acid batteries in 1859 did the recharging of batteries become feasible. The only way to charge lead–acid batteries at that time was by using Voltaic piles, which therefore became known as primary batteries, with the rechargeable lead-acid battery being termed secondary batteries. Lead–acid batteries can be recharged many, many times and are the batteries used in normal cars. Currently, most hybrid cars use a different type of rechargeable battery involving nickel–metal–hydride. These nickel–metal–hydride batteries are somewhat lighter than lead–acid ones, and in addition, their state of charge can be determined by measuring the hydrogen pressure in them. With lead–acid batteries, it is necessary to measure the specific gravity of their acid electrolyte to determine how much electric power is available.

Many different types of batteries are now known. They all have somewhat different properties. For all batteries, however, there are some basic features of importance. One of these is the energy that the battery can store compared to how much it weighs. On this scale, lead–acid batteries do not rate as high as nickel–metal–hydride batteries and are dramatically worse than lithium or lithium-ion batteries. It seems to be a rule of nature that the lighter the element, the greater its electrochemical activity is. Lithium, which is the lightest metal known, also has a high electrochemical activity. And it is electrochemical activity that determines the voltage a single battery cell can produce. Because lithium is so active, it burns rapidly with a bright colorful flame while emitting caustic oxide as smoke. One problem with lithium batteries is their tendency to catch fire. Lead doesn't burn, but lithium does. Most lithium batteries don't use metallic lithium but rather lithium salts, so the flammability problem is greatly reduced, but it is not eliminated entirely, because they need non-aqueous electrolytes.

Another basic factor is how many times the battery can be recharged before it fails. The nickel–iron battery developed by Thomas Edison still holds the record for rechargeability. Edison batteries can go through the charge–discharge cycle tens of thousands of times. Unfortunately, they also have a high self-discharge rate. Everybody who has left a flashlight (pocket torch) unused for a long time knows that batteries don't last forever even if they are not used at all. The so-called self-discharge rate for lead–acid batteries is very low. They typically lose only about one-quarter of one percent of their stored energy per day. Batteries based on nickel and cadmium, on the other hand, can lose 3% of their stored charge every day, and Edison's nickel–iron batteries lose about 10% per day. Interestingly, Thomas Edison invented the nickel–iron battery because his friend Henry Ford felt that there might not be enough gasoline available to fuel all the cars the Ford Motor Company could produce, and he was interested in another way to power them. In this thought, Henry Ford was 100 years ahead of his time.

As the cost of gasoline increases, batteries are once again being considered as an energy source for powering cars. If the flammability problems of lithium batteries can be solved somehow, they are the batteries of choice for powering vehicles, as we will see below. For one thing, they don't need to be heated up to work, as another type of high-energy battery based on sodium and sulfur does. For another, lithium batteries offer the possibility of a 200-mile (325-kilometer) driving range. One problem that remains with any battery-powered car is the long time it takes to recharge a battery compared to the time it takes to fill up with gasoline.

To see why batteries have difficulty propelling a car for long distances, note that the energy density of gasoline is about 13,000 watt-hours per kilogram. Gasoline doesn't deliver electricity, of course, but watt-hours are just another unit of energy (see Appendix II). Lead-acid batteries typically store about 40 watt-hours of energy per kilogram of their mass. The term watt-hours simply means the number of watts of power these batteries can supply for how many hours. This watt-hour value varies with the rate at which the energy in the battery is extracted. Batteries that are discharged rapidly can't deliver as many watt-hours as those that are discharged slowly.

Nickel–iron batteries, at about 50 watt-hours per kilogram, do a little better than lead–acid batteries, while nickel–metal–hydride batteries do a lot better at 90 watt-hours per kilogram, and sodium–sulfur batteries do better still at 120 watt-hours per kilogram. Typical lithium-ion batteries do the best at about 150 watt-hours per kilogram (remember that for all these batteries the watt-hour numbers depend in varying ways on the discharge rate). Even for lithium batteries, however, their energy density is only just over 1% that of gasoline.

When comparing the energy capacity of batteries, lithium-ion ones win hands down. Only sodium–sulfur batteries come close, and these batteries have to be heated up enough to melt the sulfur and sodium in order to work. Lithium batteries do not have to be heated. A closed tank can hold gasoline for years, if the gasoline cap is on tight and the tank doesn't leak. It won't self-discharge at all. But all batteries do. Also, remember that batteries need a longish time to recharge, while gasoline tanks need only minutes to refill.

One compensating factor is the fact that the energy stored in batteries can be delivered to the wheels via electric motors at a much higher efficiency than the energy in gasoline can. Internal combustion engines are heat engines and therefore are limited by Carnot efficiency considerations. Only about 25% of the energy in the tank ends up propelling vehicles. Electric motors are not Carnot limited, and 75% or more of the energy they use ends up as mechanical work. But even with this compensating factor, the gasoline tank still winds up as the lightest way of storing energy. On the other hand, if gasoline is rationed and electricity isn't, objections to the limitations of battery-powered cars with limited driving range and long recharging times will be muted.

One special class of battery deserves mention, but nobody seriously proposes it for powering cars. This class consists of nuclear "batteries." Quotation marks are used because such sources of electricity rely on radioactive decay, not chemical reactions as other batteries do. We will see in Chapter 18 how the phenomenon of thermionics can be used, together with highly radioactive isotopes, especially plutonium-238, to provide electric power. Plutonium-238 is produced in nuclear reactors, but it can only be separated from the other plutonium isotopes in used fuel rods by an enrichment process (see Chapter 5). So, it is produced instead from the isotope neptunium-237, which is also produced in nuclear reactors and is recoverable from used fuel rods by chemical means. Although not fissionable, plutonium-238 is so radioactive that it heats itself up. A sphere of plutonium-238 about the size of a grapefruit can warm itself to a dull red heat. Plutonium-238 emits helium-4 ions, which are so strongly absorbed by matter that most of them do not escape from the surface of solid plutonium. This is the isotope that was used, in combination with thermoelectric (see Chapter 6), not thermionic devices (see Chapter 18), to provide electrical power for the twin *Voyager* spacecraft, which have now flown further from the sun than any other human objects. Nuclear batteries can supply more than a million times more energy than lithium-ion batteries can, but they cannot supply sudden bursts of high power. Instead, nuclear batteries provide steady levels of electricity for incredibly long times. In the cold, inky darkness of interstellar space, nuclear batteries and nuclear reactors are about the only viable means of providing electrical power to small spacecraft.

In addition to liquid fuels, chemical batteries, and compressed air, there is yet another way of storing energy in cars, and that is by using flywheels, as discussed next.

Flywheels

Flywheels have long been considered as a means for powering cars, but this application is fraught with problems. Flywheels, whose engineering is entirely different from that of compressed-air systems or batteries, can be effective in delivering large amounts of stored energy fairly rapidly, but such applications for flywheels are specialized. The amount of energy that can be stored by flywheels is orders of magnitude less than that for either pumped-water or compressed-air systems.

The basic idea of a flywheel is to store energy in the form of the kinetic energy of motion. Everyone realizes that an asteroid moving at a high velocity releases a lot of energy when it hits the Earth. The energy released on impact is just the energy the asteroid had because it was in high-speed motion. A crashing car does the same thing, that is, it releases its kinetic energy of motion on impact. An important fact is that the kinetic energy of any moving object increases as the square of its velocity. How can kinetic energy be used to store energy? The answer is simple: keep something moving in a circle. This is the principle of a flywheel. The energy of a flywheel increases as the square of its angular velocity, that is, the rate at which it is spinning. To store lots of energy, flywheels have to be large and rotate very rapidly.

One problem with flywheels is that they continuously lose some of their energy because of air friction, unless they are spinning in a vacuum. Naturally, building a huge vacuum chamber is expensive. To avoid the need for a large vacuum chamber, large flywheels normally store energy for only relatively short periods of time, unlike pumped-water systems, whose energy-storage times can be very long. As an example of flywheel energy storage, consider the very large flywheel that was built to supply pulses of power to the National Magnet Laboratory that used to exist at the Massachusetts Institute of Technology in Cambridge, Massachusetts (the National High Magnetic Field Laboratory is now located at Florida State University, Tallahassee, Florida). It takes a lot of electric power to drive an electromagnet hard enough to produce magnetic fields 500,000 times stronger than that of the Earth, and the National Magnetic Laboratory would need to suddenly draw 32 MW of power from the Cambridge electric grid when one of its powerful magnets was turned on, which would have dimmed lights all over town. To prevent this, a very large flywheel system was built. This massive flywheel, as usual, was made from cast iron. This flywheel system could be brought up to speed by slowly drawing power from the Cambridge electric power grid after midnight. Once set spinning, this flywheel could supply, for short times, the 32 MW needed to ramp magnetic fields up to full strength without dimming lights in Cambridge.

For megawatt-seconds of energy, flywheels are the way to go, but for storing megawatt-days of energy, it turns out they are impractical. They are not very practical for powering cars, either. To begin with, flywheels are giant gyroscopes that can interfere with the way a car handles, as anyone who has played with a gyroscope knows. The energy stored in flywheels can be released explosively if its bearings or the flywheel itself fails. Finally, flywheels, even those revolving at enormous speeds, can't power a car over the distances that batteries or compressed air can. Consequently, flywheels have been virtually abandoned for the purpose of powering cars. They still have applications for supplying pulses of electric power, but even in these other applications they have competition.

Capacitors can also be used to supply large or small pulses of electric power, so let's consider them next.

Capacitors and Dielectrics

Capacitors deliver the energy they store even more rapidly than flywheels, because of the way they work. Capacitors use electromagnetic forces, not the kinetic energy of motion, to store energy. The world's very first capacitors were called Leyden jars because they came into use in the Dutch city of Leyden around 1745. In those days, the big scientific development was static electricity, and people were interested in sparks. Leyden jars enabled an electric charge to be stored, and they could "capture a spark," as Benjamin Franklin is reported to have done in his famous kite experiment. The first Leyden jars were just stoppered glass jars of water with a connection to the water inside via a nail driven through the stopper. The function of the glass was to act as a dielectric material, that is, a material that lets the electric field pass through but blocks the flow of electric current. A dielectric is really an insulator.

It had been discovered that static electricity could be stored in such jars and then released by touching the nail. It wasn't long before the water was replaced by thin layers of metal, such as gold foil, one inside the glass jar and one outside. The principal here is that two electrical conductors, such as metal foils, separated by an insulator can retain electrostatic charges. They work because positive and negative electric charges attract each other through the insulating dielectric material and are held in place by their mutual attraction. All capacitors still use this principle, and capacitors are made by separating two conductors by a dielectric. The amount of energy that a capacitor can store depends on its size and, most importantly, on the square of the voltage used to charge it. The maximum voltage that can be used is set by the dielectric strength of the insulating material used. Dielectric strength is just a measure of the voltage at which the insulator breaks down and a capacitor shorts out. One additional factor is that some insulators are better than others because they can increase the magnitude of the electrical charge that the capacitor can hold. How can they do this? The answer is that the atoms of some insulators can shift around even though they are still held in place. Shifting charges around results in an increase in the total stored electrical charge and enhances the energy-storage ability of the capacitor. Work has also been done on using charged ions in a liquid (ions are atoms with either missing or extra electrons) to improve the ability of capacitors to hold an electric charge. Charged ions sound exotic, but simply dissolving common salt into water produces charged particles in the form of sodium ions (sodium atoms with a missing electron) and chlorine ions (with an extra electron). The ocean is chock full of ions. Remember that the first Leyden jars were just water in a bottle. They might have worked better if they had held salt water instead of fresh water.

Years of engineering improvements have dramatically increased the ability of capacitors to store energy, but even so, they are still not practical for applications on the scale of pumped-hydropower or compressed-air energy storage. Nobody seriously suggests using capacitors as the sole means of powering cars. However, it has been suggested that capacitors could be used to supply sudden bursts of power to hybrid vehicles in addition to what its batteries can supply. In particular, capacitors using ions have been suggested for this purpose. How practical this ion approach will be in improving hybrid vehicles remains to be seen.

Capacitors do have one very great advantage over other methods of storing energy: they can release the energy they store extremely rapidly. As an example, consider lithotripter (from the Greek for stone-breaking) machines used to treat kidney stone disease. These devices produce an acoustic shock wave, which is focused to fragment kidney stones (which have the mechanical properties of chalk), no surgery required. To create the needed acoustic shock wave, capacitors holding about 10 joules of energy can be discharged across an underwater spark plug to create a sudden acoustic pulse. Ten joules is not much. It's just the energy it takes to run a 10-watt light bulb for one second. But, and this is the important point, those 10 joules of energy are delivered in about a microsecond. Power, of course is energy per unit of time, so 10 joules delivered in one microsecond gives a power level of 10 million watts. That's a lot of power, and capacitors are by far the most practical way of delivering extremely high levels of power for very short times, such as microseconds. But they are not good at storing a lot of energy. The capacitors needed to store 10 joules weigh about 10 pounds (4.5 kg) and operate at 20,000 volts. The fact that capacitors deliver the energy they store with extreme rapidity is the primary reason they are used for energy storage. Capacitors are also routinely used in building electrical circuits of all kinds, and in such applications they usually store energy on a micro-scale. At the other end of capacitor sizes, a bank of enormous capacitors at the Los Alamos National Laboratory is used to pulse-power a magnet for generating magnetic fields two million times stronger than that of the Earth, a level that generates magnetic effects that cannot otherwise be studied in detail. The power levels these magnets need are truly colossal, and banks of large capacitors, each as big as a refrigerator, are the only practical way to achieve them, even if these giant capacitors can supply current for only a few microseconds. Because capacitors don't hold a lot of energy but can deliver vast amounts of power for very limited times, they are of no use all by themselves for powering cars. Inductors, on the other hand, can easily release their stored energy more slowly and controllably, so let's consider them next.

Inductors: Storing Energy with Magnetic Fields

Just as it takes a lot of energy to generate a powerful magnetic field, magnetic fields also store a lot of energy. Michael Faraday (1791–1867), the same man who discovered that moving a copper wire in a magnetic field produces an electric current in the wire, was the discoverer of this effect. Basically, inductors are just coils of wire that can also be electromagnets. A coil of wire with an electric current running through it produces a magnetic field. The energy stored in this magnetic field can be recovered, at least partially, as the level of electric current through the coil is reduced. For normal conductors a lot of the energy is wasted in the heat generated by the electrical resistance of the wire in the coil. For superconductors, of course, there is no electrical resistance, and so there is no conversion of stored energy into heat via electrical resistance. The point is this: the more coils of wire an inductor has, the greater the inductance of a wire coil and, consequently, the greater amount of energy it can store. Not so obvious is the fact that the energy stored in a coil of superconducting wire increases as the square of the current it is carrying. Because of these two facts, large coils of superconducting wire carrying thousands of amperes store enormous amounts of energy. Superconducting energy-storage systems the size of houses can store enough energy to power thousands of homes for days. Capacitor banks of similar dimensions can't come even close to this.

One problem, of course, is that the superconducting coil is kept at temperatures around that of liquid helium, $4.2\,°K$ ($-268.8\,°C$ or $-451.8\,°F$). Capacitors have the advantage that they don't need to be cooled to cryogenic temperatures. If a super-conducting coil warms up very rapidly, the energy it stores may be released explosively. Even so, magnetic energy storage using superconductors has been seriously considered for load-leveling applications, just like pumped-hydropower energy storage. Of course, pumped-hydropower systems don't explode if they warm up, and superconducting load-leveling systems are not yet in commercial service. But superconducting, load-leveling systems could be more efficient than pumped-hydropower systems, because the only energy loss is that needed to keep the wire coils cold. Also, they don't need a nearby mountain and a lake to store large amounts of water.

It is important to remember that energy can be stored only after it is generated, so storage by any method whatsoever is not a solution to the problem of the world's ever-growing need for energy. But energy storage can, and does, play a role in the effective use of the energy-generation facilities that humanity already possesses. In particular, the engineering of reliable, long-life batteries based on the very light element lithium appears to offer the best near-term possibility for making electric vehicles practical for local transportation. The sooner battery-powered cars can be brought to the mass market the better. The same is true of fuel cell cars, as discussed in Chapter 16.

Non-fossil fuels also will have a major role to play in keeping transportation going as the age of petroleum slowly winds down, so we'll look at them in the next chapter.

Recommended Reading

Andrews, John. *Energy Science: Principles, Technologies, and Impacts*. New York: Oxford University Press, 2007.

Gladwell, John S. and Calvin C. Warnick. *Low-Head Hydro: An Examination of an Alternative Energy Source*. Moscow, ID: Idaho Water Resources Research Institute, 1978.

Seeber, B. and David Larbalestier. *Handbook of Applied Superconductivity*. London: Macmillan Journals, 1998.

Sheahen, Thomas P. *Introduction to High Temperature Superconductivity*. New York: Plenum, 1994.

12
Green Fuel: Biodiesel, Alcohol, and Biomass

Heap on more wood! – the wind is chill.

Sir Walter Scott, *Marmion*, 1808

Because petroleum is just a product of long-dead plants, it is obvious that plant matter can be used to produce oil. Making alcohol, biodiesel, and other fuels, including gasoline itself, from biomass is one long-term approach to solving the world's need for vehicle fuels. One caveat: current world petroleum consumption is over 85 million barrels per day. It might take more than 20 barrels of biomass to produce one barrel of fuel, not even counting the fuel it takes to harvest and process this biomass. Replacing gasoline with fuel from biomass is a monumental task. Worse yet, the world's consumption of petroleum is increasing relentlessly. Even so, green fuel offers both great possibilities and great problems. Let's look backwards briefly to see how nature has produced the coal, petroleum, and natural gas we have been using at an ever-increasing rate for the last century.

Since time immemorial, animals have been breathing in oxygen and breathing out carbon dioxide, while plants have been doing the opposite. Talk about a renewable system! After a very long time, human beings started burning wood to keep warm and cook food. A large numbers of us still do. That's how our current troubles first started, as the Greek poet Hesiod wrote long ago in the couplet that opens Chapter 1. Even now, more than half of all biomass consumption involves burning wood.

Fossil fuels such as coal are just the remains of plants long dead and gone. In some sense, we are still in the plant-burning business. Biomass simply means plant matter. Biofuels are chemically altered forms of the hydrocarbons that plants produce from sunshine, carbon dioxide, and water. Alcohol is one example. Biodiesel is another. With enough chemical engineering it is even possible to produce gasoline itself from sugar. The potential for producing renewable energy from vegetation was never in doubt. What is in doubt is whether or not this energy resource is sufficient to raise living standards for the whole planet. The answer could be yes, if the planetary population were not as large as it is and if demands for energy were not so huge and rising so fast. But the global population and world energy use are both great enough that biomass and biofuels, like so many other

Energy Demand and Climate Change: Issues and Resolutions. Franklin Hadley Cocks
© 2009 WILEY-VCH Verlag GmbH & Co. KGaA, Weinheim
ISBN: 978-3-527-32446-0

energy sources, can't do more than supply a fraction of what is needed to keep things going as they are without dramatic changes.

As the world's population increases, it's going to take absolutely every possible energy source to increase living standards around the globe. Biomass and biofuels can safely be predicted to play an important role, provided that the planetary population does not increase so much that all usable land must be used for food production. Even if land must be used for growing food, after every harvest a lot of plant material is left lying on the ground. At least some of it could be used for biomass or biofuel. However, if all of it were used and none returned to the soil, it might not be long before crops did not grow very well. The amount of biomass that must be returned to the soil to ensure crop production probably needs to be studied in even greater detail than it has been. Of course, the industrial-style production of organic matter from algae can avoid or at least reduce the need to use large amounts of agriculturally productive land for fuel production. Algae, properly nourished, can produce 10 times more biofuel per acre or hectare than agricultural food plants can. Some forms of algae can grow in brackish water or seawater. Using ocean water or brackish water to grow algae does not reduce the land area available for growing crops. One major problem, of course, is the cost of producing biofuel from algae. This cost might or might not be higher than that of producing biodiesel from plant oils or alcohol from cornstarch (see Chapter 20 for more about ocean algae and using large-scale blooms of ocean algae to reduce atmospheric carbon dioxide).

In addition to turning plant matter into fuel, it is naturally possible to simply burn it to produce electricity as well as heat. This process is easily applied and is already underway, so let's look at this possibility first. District heating, in which a central plant supplies hot water or steam to several buildings, can be powered by burning wood if an ample wood supply is locally available. District heating can, and in a few places already does, make use of the waste heat from small electric power plants fueled with wood. In some district heating operations, the heat is supplied renewably by using biomass. With the addition of cooling systems such as ammonia and water, which are used in natural gas–powered refrigerators, or the newer lithium bromide and water refrigeration technique, the same piping that supplies warm water in winter could also supply chilled water in summer, but facilities that can do both are currently only of prototype sizes.

An uncountable number of dwellings on all continents use wood for heating. As the cost of oil and coal increases, it is expected that both residential and non-residential use of wood and biomass for heating will expand, especially for school heating. This approach to supplying fuel for heating is more difficult in large-population centers. But it is increasingly attractive in rural areas, where schools tend to be small, houses dispersed, and trees numerous. In some places biomass is already being used for producing both electric power and heat. In Vienna, a biomass-fueled combined heat and power facility supplies 23 megawatts of electric power in the summer. In the winter, this facility supplies 15 megawatts of electric power and 37 megawatts of thermal power to thousands of households. Many other, smaller facilities like this also exist.

Central steam plants, mostly coal-fired, heat many universities as well as business or industrial complexes and portions of some cities and towns worldwide. Fueling any central heating, cooling, or electric plant with biomass requires an ample supply of wood. It takes about 200,000 tons of wood to keep a two-megawatt power plant going for a year. However, logging and trucking wood to the power plant currently relies on fossil fuel, so such a system is not entirely renewable. Of course, horses, wagons, and handsaws could be used to keep the wood supply coming, but doing so somehow seems a step backward in time. One advantage of wood-fueled heating, lighting, and cooling systems is that a lot of the money for fuel is paid to the local working population. Biomass-fueled district heating systems are usually small because otherwise the cost of transporting the needed wood or other biomass becomes prohibitive as the size of required woodland gets larger.

Naturally, the use of wood and other burnable plant matter or animal waste varies a lot from country to country. The least developed countries tend to supply a far greater percentage of their annual energy needs from biomass. But, of course, their per capita energy budgets are also lower. In the United States, for example, only about 3% of the national energy supply comes from biomass. Almost all of it is used for heating. One trouble with biomass is its low concentration compared, for example, with coal. Coal is usually concentrated in veins. In contrast, biomass is usually spread out over the surface of soil or water. Of course, there are special situations, and pulp and paper plants are one of these. In North America, many pulp mills involved in paper production burn waste, such as bark, to generate electricity. This waste has to be disposed of somehow anyway, and burning it to produce electric power makes economic sense. Sugarcane mills can do the same thing with their waste biomass.

In recent years there has been increasing interest in using the ethyl alcohol produced by fermentation of sugar and starch from corn, sugarcane, or sugar beets to power cars and as an additive to gasoline. Part of this interest may be politically driven, because of the economic boost greater demand brings to farming regions by increasing corn, sugarcane, or sugar beet prices. Corn is used in foods, so boosting its price ends up inflating the cost of many grocery products. But corn is certainly not the only crop that can be used to produce alcohol. In fact, a lot of other crops, especially sugarcane or sugar beets, can produce more alcohol per acre or hectare. Brazil, in particular, has succeeded in replacing a substantial part of its gasoline needs with ethanol from sugarcane. But even with sugar- and carbohydrate-rich crops like corn, sugarcane, and sugar beets, most of the plant material does not end up as alcohol, and a lot of cellulose-rich stalks and waste are left behind. Alcohol can be made from cellulose, but it's not as easy as making it from sugar or starch.

It is important to realize that cellulose is nothing more than polymerized sugar. The stomach juices of cows and other grass-eating animals contain enzymes that break down the polymerized sugars that compose cellulose. Once they are broken down, the resulting unpolymerized sugars can be absorbed into the bloodstreams of grass-eating animals just as the sugar (glucose) produced by human digestive processes is absorbed into our bloodstreams. If the special enzymes that break polymerized sugar apart can be produced cheaply, cellulose plant material can be used to produce alcohol in vast amounts. Cellulose can produce more alcohol than the sugar and the starch squeezed out of corn, sugarcane, or sugar beets. The enzymes necessary to decompose the polymerized sugar that makes up cellulose can be produced now, but their cost and complexity have so far prevented widespread production of alcohol from cellulose. Because cellulose is not human food, its use in producing alcohol for vehicles may not increase the price of food very much. There is a lot that is appealing about turning yard waste into fuel for cars. It is already possible to turn wood into methyl alcohol (methanol) by pyrolysis (thermal decomposition without combustion), but this process does not yield as much alcohol from a given amount of wood as enzymatic depolymerization followed by fermentation could.

Besides producing the liquid fuel alcohol, the gaseous fuel methane can also be made from biomass. Natural gas is produced by the bacterial decay of plant matter. Bacteria can eat either the sugar or the polymerized sugar in plant cellulose to produce methane. In swamps, bacteria-produced methane sometimes bubbles up in sufficient quantity to make a combustible mixture with air. It is a rare sight, but slowly burning dilute air–methane mixtures floating around may give rise to reports of will-o'-the-wisp ghostly sightings in swamps at night. The interaction of bacteria with animal or human wastes can also produce methane, and industrial-scale pig farms are being pressured to use their lagoons of pig excrement to produce methane. Because methane does not contain nitrogen, the matter left over once methane production is completed is rich in water-soluble nitrogen and makes a good fertilizer. So far, the cost of such methane far exceeds that of natural gas, but this may not always be so.

Biodiesel

Diesel engines can burn plant oils directly if they have to. A diesel car will run on plain peanut oil (producing an exhaust redolent of peanuts), provided it is started up with regular diesel fuel. The engine also runs better if the peanut oil is diluted a bit with regular petroleum-based diesel fuel. During World War II, when diesel fuel was hard to obtain, plant oils were widely used to power vehicles in Africa and Asia. Biodiesel is diesel fuel produced by the chemical engineering of plant oils, so that no normal diesel oil is needed to start or run a diesel engine. One reason why plant oils need to be modified to approximate diesel fuel from petroleum is that plant oils contain glycerin. Glycerin is a lot more viscous than normal diesel oil. To understand viscosity, it is helpful to know its linguistic history. Viscosity actually means "mistletoe-ishness." *Viscum* is the Latin name for mistletoe, that remarkable plant that grows on trees. The seeds of mistletoe plants contain thick, gooey syrup that causes these seeds to stick when birds eating them clean their beaks by rubbing against tree branches. High-viscosity fuel is not good for diesel engines, and the glycerin in plant oils is removed to produce biodiesel.

Raw plant oils are more viscous than diesel oil, but glycerin can be taken out of plant oils via straightforward chemical engineering, thereby making them less viscous. It is also possible to imagine making biodiesel on a large scale, using algae as the basic feedstock. Biodiesel can be easily transported over long distances using the same pipeline network that is now used to transport oil, while alcohol can't, at least not as easily. It turns out that alcohol causes more corrosive problems with steel than oil and biodiesel do, in part because ethanol almost always contains some water. Using existing pipeline networks to transport alcohol could lead to major corrosion problems, but using normal steel pipelines to transport biodiesel wouldn't. This fact means that remote, unpopulated, semi-arid or arid regions, which can't be used to grow normal crops, might be used to grow algae or drought-resistant shrubs yielding plant oil. If algae is grown in closed systems, the water needed is greatly reduced compared to that needed for normal farming. Algae can even grow in oxygen-starved lakes. The accumulation of dead masses of *lacustrine* algae forms an organic mixture of so-called kerogen in sedimentary rock that can be heated to yield heavy oils. Kerogen deposits of various kinds are part of the rocks that form oil shale. Without waiting for nature's rock-forming process, algae can be turned into biodiesel on a large scale once huge quantities of it are available. The biodiesel approach to vehicle fuel production might not compete with food production in the way that producing ethyl alcohol from corn does.

Certain shrubs, such as *Jatropha curcas*, produce non-edible, bean-like seeds that are very oil rich. These beans can be crushed to produce as much as 35% by weight of plant oil. *Jatropha curcas* is a perennial plant that can thrive on as little as 10 inches (25 centimeters) of rain per year, and it grows well in what would otherwise be scrub land This plant does not need artificial insecticides, as it produces its own insect poisons. Large-scale cultivation of *Jatropha curcas* has already begun, but so far its contribution to the world's supply of diesel fuels is negligible. There may be other plants that have similar properties. Just how far plant breeding can be used to expand the oil-producing properties of *Jatropha curcas* is virtually unexplored. The possibilities for producing diesel fuels from plant oils are many and varied.

Diesel fuel from either plant oils or petroleum naturally requires a diesel engine, which is quite different from a gasoline engine. Rudolph Diesel (1858–1913) was a German engineer who achieved his dream of improving the efficiency of internal combustion engines. He accomplished this feat by building an engine that operated at higher pressures and temperatures than gasoline engines. Diesel's engines must burn something less volatile than gasoline. They don't need spark plugs to ignite fuel. The temperatures reached during the compression of the fuel–air mixture in diesel engines are high enough to cause ignition without the need for an electrical spark, although non-sparking but hot "glow plugs" help start diesel engines, especially in cold weather.

To this day, diesel engines can get better mileage than gasoline engines, due to their higher operating temperatures. It is also simpler to turn plant oil into biodiesel than to make alcohol from plant sugars, although both processes are relatively straightforward. If cellulose rather than plant sugar is used to make alcohol, then an acre or a hectare of farmland could end up producing far more alcohol than biodiesel. Alcohol is utterly unsuited to diesel engines, but it is well suited to gasoline engines, with a little carburetor modification. Many of Henry Ford's first cars were fitted with adjustable carburetors for just this purpose. In his day, gasoline seemed limited in its future supply, and in this thought he was right.

Rudolf Diesel's engine was a true engineering advance, and for a while he seemed to be profiting a lot from what he had accomplished. But he died nearly broke because his engineering ability was much better than his investment acumen (he invested heavily in Balkan oil schemes, for example). He was 55 years old in 1913 when he jumped, fell, or was pushed overboard one night from the steamship *Dresden* on his way to England to persuade the British Navy to use diesels to power submarines, which they soon decided to do anyway. Diesel knew full well that plant oils could be used in his new engines, and he felt that the use of plant oils rather than petroleum to fuel them would give a big boost to agriculture. In this thought he was another prophet ahead of his time.

Renewable biomass or biodiesel fuels are not practical for 1000-MW power plants. It would take a forest of several thousand square miles of pine trees to keep a 1000-MW power plant continuously fueled with wood. Biofuels and biomass have an important role to play, but biomass seems to be best when used in small electrical power facilities, not giant power plants. Producing electricity using biomass or biofuel fuel is more labor intensive than using coal or nuclear energy. However, the carbon footprints of power from biomass and biofuel are far smaller than that from coal and oil, although not as small as that from nuclear power plants.

An increasing level of carbon dioxide affects the growth of plants. Vines grow especially well as carbon dioxide levels rise, thereby boosting biomass production. All plants need water, and, surprisingly, the evaporation of water from plant leaves slows down as the carbon dioxide level of the air goes up. It turns out that leaves contain tiny pores, called stomata, that shrink in size as the level of carbon dioxide increases. The result seems to be that if the carbon dioxide level of the atmosphere doubles, as it appears likely to do, the rate of water evaporation from plant leaves will go down by about 6%. This may not seem like much, but there are a lot of leaves and trees around. Less water loss from leaves means more of it is left in soil as the transport of water from plant roots to leaves slows down. Higher levels of soil moisture in forests could lead to more flooding after heavy rains. Greater levels of carbon dioxide in the air leads to warmer global temperatures and an increased rate of evaporation from the oceans, leading to more frequent and heavier rainstorms and more flooding problems, as well as higher average humidity. The ecosystem of the world is very complex, indeed.

The fraction of total energy supplied by biomass has been steadily decreasing ever since the advent of the industrial revolution in the 18th century. But every energy source counts, and biomass and biofuels have the great advantage of being renewable. Burning them mostly just returns to the air the CO_2 plants use to grow. In 1900 about 35% of the world's energy came from biomass, primarily wood. In the next 100 years the world's total energy usage grew by more than a factor of 10, but the bulk of this increase came from burning fossil fuels, not biomass. In the year 2000, the fraction of the world's total energy from biomass or biofuels was down to about 10%. If world energy usage increases by another factor of 10 over the course of the 21st century, it seems likely that the traditional use of biomass will supply only a modest fraction of it. But that situation could change if massive switching to biofuels is achieved.

Biomass and biofuels have an important role to play in humanity's seemingly ceaseless drive for more energy, but don't count on them to supply the fuel needed to remake the world anytime soon. One fundamental difficulty that all biomass and biofuel plans must face is the fact that plants use only about 3% of the energy in sunlight. Solar cells can convert more than 30% of the energy in sunlight into electricity. Of course, plants grow and reproduce themselves; solar cells have to be manufactured. Theoretically, around 11% of the energy in sunlight could be used in photosynthesis, but there must be some fundamental reasons why plants don't reach this level of sunlight usage. Chlorophyll, the compound that plays such a critical role in plant life, may not be the best such compound for this purpose, but it's the one that plants have developed. Interestingly, magnesium plays a role in chlorophyll analogous to the role that iron plays in hemoglobin. That's one reason why people eat green vegetables – to get the magnesium that we need to live. It may be possible via miracles of chemical engineering to produce artificial photosynthesis processes that use more than 3% of the energy of sunlight. It also may be possible to engineer plants genetically to use something better than chlorophyll for photosynthesis and to produce plants like kudzu on steroids. Kudzu, a vine-like plant that was imported from Asia to North America, seems to outgrow most native North American plants. Artificially generated super plants would bring their own set of problems, naturally, but they could also produce a lot more biomass and biofuel than normal plants.

Plants such as water hyacinths and algae already generate biomass faster than most other plants. The oceans offer the possibility of large-scale production of the salt-water, gene-altered equivalent of water hyacinths and algae. Harvesting such a resource from the oceans and drying it requires a lot of effort and energy. Biomass and biofuel are important, and the possibilities they offer should be exploited, along with all other energy resources. Maintaining advanced civilization over the coming centuries is not going to be easy. Time to begin major efforts, both public and private, to replace petroleum with fuels that can be produced quickly from plants is quickly running out. A multifaceted effort to accomplish this result should be started now. Producing methanol from corn starch is only the faint beginning of this process.

Recommended Reading

Kaltschmidt, Martin, Wolfgang Streicher and Andreas Wiese, eds. *Renewable Energy: Technology, Economics and Environment*. Berlin: Springer, 2007.

Melis, Anastasios and Thomas Happe. "Hydrogen Production: Green Algae as a Source of Energy." *Plant Physiology* 127 (2001): 740–748. See also: http://www.plantphysiol.org.

Pahl, Greg. *Biodiesel: Growing a New Energy Economy*. White River Junction, VT: Chelsea Green, 2005.

Wilson, John R. and Griffin Burgh. *Energizing our Future: Rational Choices for the 21st Century*. Hoboken, NJ: Wiley-Interscience, 2008.

Part III Dreams

Energy Demand and Climate Change: Issues and Resolutions. Franklin Hadley Cocks
© 2009 WILEY-VCH Verlag GmbH & Co. KGaA, Weinheim
ISBN: 978-3-527-32446-0

Introduction

Oh my soul, do not long for miracles, but rather be content to
exhaust the limits of the possible.

Pindar, *Pythian Ode 3*, 474 B.C.

Part II reviewed technologies that can be implemented immediately, above and
beyond the existing and well-known sources of energy that fossil and nuclear fuels
represent. Part III presents new possibilities, some of which have only begun to
be developed. They are all technically difficult but offer major energy rewards if
they can be fulfilled. All dreams are not the same, of course, and some of the possibilities presented in Part III are much easier to bring into being than others. The
technologies presented first are much closer to being able to be used on a large
scale than are the ones discussed later on. Breeding nuclear fuel and magnetohydrodynamic power generation have both already been demonstrated in practice.
Space elevators have not. All are scientifically possible. None is free of pitfalls or
the possibility that they will prove to be unacceptably difficult or complex. Imple-
menting some of them, or even just a few, would help greatly in dealing with our
potential energy Waterloo. But doing so will not be easy or cheap. The sooner we
determine which methods can eventually contribute towards solving humanity's
energy and climate problem and which can't, the better chance there is for bringing the successful ones into service while there is still time.

Energy Demand and Climate Change: Issues and Resolutions. Franklin Hadley Cocks
© 2009 WILEY-VCH Verlag GmbH & Co. KGaA, Weinheim
ISBN: 978-3-527-32446-0

13
Breeding Nuclear Fuel

> *Should a scheme be devised for converting to energy even*
> *as much as a few percent of the matter of some common*
> *material, civilization would have the means to commit suicide*
> *at will.*

Report on the Development of the Atomic Bomb, 1945

> *The atomic age has moved forward at such a pace that every*
> *citizen of the world should have some comprehension, at least*
> *in comparative terms, of the extent of this development, of the*
> *utmost significance to everyone of us.*

President Dwight David Eisenhower, *Atoms for Peace*
address to the United Nations, December, 1953

Like almost all technical advances, nuclear technology has applications for peace or war, and the worldwide use of atomic energy for peaceful purposes is expanding. "Breeding" nuclear fuel is a way of multiplying by a factor of more than 100 the amount of fuel for nuclear fission power plants. It is the process of converting non-fissionable isotopes into the fissionable ones required for nuclear power reactors. Breeding allows far more of the world's uranium to be used for generating power than is now the case. In addition, breeding allows thorium to be turned into a usable nuclear fuel for reactors. Named for Thor, the Scandinavian god of war, thorium has only one natural isotope, thorium-232. This isotope cannot be used directly for fuel for nuclear power plants because it isn't fissionable. But thorium-232 can be bred into fissionable uranium-233. This possibility offers up a nuclear fuel energy resource three times more abundant than natural uranium.

The masses of the nuclear fragments that result from fission don't add up to the original mass because some of this mass has been converted into energy. But reactors convert only about 0.1% of the mass of each fissioning atom into energy. That is not a large percentage, and initially most of the energy from a nuclear reactor comes from the fission of uranium-235 atoms. Only 0.714% of natural uranium is uranium-235. The rest is uranium-238 (98.28% with a tiny amount of uranium-234 (0.006%). Turning the 99.28% of uranium that is uranium-238 into fissionable plutonium-239, by exposing this uranium to a flux of fast neutrons,

Energy Demand and Climate Change: Issues and Resolutions. Franklin Hadley Cocks
© 2009 WILEY-VCH Verlag GmbH & Co. KGaA, Weinheim
ISBN: 978-3-527-32446-0

makes most of the world's uranium usable for generating electric power, not just the 0.714% that is uranium-235. How marvelous it would be if, rather than being an invitation to suicide, making a larger percentage of uranium convertible into energy would help civilization survive for a long time. Breeding nuclear fuel makes this result possible.

All reactors produce fissionable fuel by converting uranium-238 to plutonium-239, but normal reactors produce less fuel than they use up. In fast breeder reactors, more fissionable plutonium-239 is produced than the amount of uranium-235 consumed. Of course, there are problems, not the least of which is the very high neutron radiation levels required. In addition, nuclear weapons can be made using the plutonium-239 produced by breeder reactors. This plutonium (Pu-239) is extractable by chemical processing from used fuel rods. If removed after the reactor has been operated for only a short time, say a week, it does not have to be enriched to yield weapons-grade material, because the concentration of plutonium-239 versus its other isotopes at that point is high enough to make bombs. If a reactor is operated for a long time, such as six months, before the plutonium is extracted, other isotopes of plutonium besides plutonium-239 have accumulated. These other isotopes degrade the plutonium so that it is no longer rich enough in plutonium-239 to make weapons. However, such non-weapons-grade plutonium is still useful as a fuel for nuclear reactor operation. Complicated isotope separation rather than relatively simple chemical processes is necessary to produce weapons-grade Pu-239 from the mixture of plutonium isotopes present in reactor fuel after the reactor has operated for an extended time.

Fast Breeder Reactors

The special type of reactor needed to turn uranium-238 into plutonium-239 is called a fast breeder reactor. All neutrons produced by nuclear fission initially move very fast. For normal reactor operation, these neutrons are slowed down, that is, moderated, usually by repeatedly bouncing off the hydrogen atoms in water in order to better fission uranium-235. In a fast breeder reactor, neutrons are not moderated in order to interact better with uranium-238 and make uranium-239. How do they do this? It works this way: high-energy neutrons from the fissioning of uranium-235 (or from fissioning plutonium-239) merge into uranium-238, which emits an electron to make neptunium-239, which in turn emits another electron to produce plutonium-239. Every electron emission from the nucleus converts a neutron into a proton, thereby increasing the atomic number. The breeding ratio is the amount of fissionable fuel produced divided by the amount consumed. This turns out to be the ratio of the isotope plutonium-239 produced divided by the amount of uranium-235 or plutonium-239 consumed. Currently, the best breeding ratios obtained are about 1.2 to 1, but a ratio of 1.4 to 1 appears to be theoretically possible.

The core of a fast breeder reactor is made up of either plutonium fuel rods or uranium fuel rods enriched to over 15% in fissionable material so that the fissioning rate, and hence the neutron flux, is very high. This 15% enrichment is five times higher than the enrichment level used in pressurized-water power reactors. The reactor core, fueled with either uranium-235 or plutonium-239, is surrounded by a jacket of "fertile" uranium-238. Each time an atom of plutonium-239 or uranium-235 fissions, it gives off heat and around two or three neutrons. Some of these neutrons are captured by the uranium-238. Each of the neutrons captured by an atom of uranium-238 produces an atom of plutonium-239.

Accomplishing a high breeding ratio is not easy. For one thing, fast breeders need to be cooled by something other than water to avoid moderating the neutrons generated by uranium-235 fission. For this purpose, liquid sodium can be used, and it is usually pumped continuously through the reactor. But molten sodium burns with dramatic rapidity in air. Fast breeder reactors need uranium that has been enriched with uranium-235 to a level of over 15% in order to achieve the neutron levels needed to attain a high breeding ratio like 1.2 to 1.

In 1966 the fast breeder reactor Fermi I, located near Detroit, Michigan, effectively self-destructed, but without loss of life. It seems that a piece of zirconium metal came loose, possibly by liquid-metal embrittlement from the liquid sodium, and blocked the flow of liquid sodium cooling the reactor core. The resulting temperature of the highly enriched fuel then increased enough to cause meltdown. The French-built Phoenix and Super Phoenix breeder reactors were shut down, too, but only after several years of safe operation. At the present time, it turns out that the cost of mining fresh supplies of natural uranium and enriching it so that it can be used in light-water (normal water) reactors is cheaper than making nuclear fuel in a breeder reactor, but this will not always be so.

One problem with early designs for breeder reactors was the swelling of the stainless steel in them due to radiation damage from the high neutron flux, but this swelling problem is minimized with suitable reactor engineering and design. It also deserves mention that the small experimental breeder reactor EBR-II operated successfully for about 30 years at the reactor test facility in Idaho Falls, Idaho. This small, 20-megawatt reactor was built and operated to demonstrate that a liquid sodium–cooled breeder reactor could be designed in a fail-safe manner such that the reactor core would not melt down. In this passively safe design, a pool of molten sodium was used in addition to circulating pumps that forced the liquid sodium coolant through the core. Natural convection in the pool of liquid sodium can cool the core without pumps, and this natural convection increases as the core temperature rises. In addition, the fuel elements were designed so that, as they heated up, their natural thermal expansion decreased the rate at which nuclear fission occurs. The inherent safety of this design was demonstrated by shutting down the coolant pumps and observing that the reactor self-stabilized to near its normal operating level.

It appears that as long as there is enough uranium-235 available at a reasonable price, fast breeder reactors will not be needed, unless national self-sufficiency becomes a driving force. But in the long term, breeder reactors offer the possibility of dramatically increasing the amount of fissionable nuclear fuel available to humanity. Breeder reactors are a technology that is clearly possible, but to become practical a lot of engineering and testing will be needed, and this development work will be costly. At the moment it is cheaper to produce more uranium from lower-grade ore than to breed uranium-238 into plutonium-239. In addition, the technical problems with liquid sodium are serious, and thus considerable effort has gone into analyzing the use of molten salts instead. However, a problem with molten salts is that they are not as good as liquid sodium is for removing heat, but at least molten salts don't burn.

So far we have been discussing breeding the fertile but unfissionable isotope uranium-238 into the fissionable isotope plutonium-239. As mentioned in the opening paragraph of this chapter, there is another fertile but unfissionable isotope, thorium-232. This isotope, which happens to be the only natural isotope of thorium, can be converted by breeding into fissionable uranium-233. Thorium-232 offers a nuclear energy resource much more abundant than uranium. Breeding thorium into uranium-233 requires a reactor called a thermal breeder, whose design is different from that of the reactor needed to breed uranium into plutonium. Thermal breeder reactors will be considered right after we look at the Clinch River Breeder Reactor project.

Clinch River Breeder Reactor Project

The Clinch River reactor was to be a large fast breeder reactor. During the 1950s and 1960s, there was strong support for nuclear power. At that time, it was thought that nuclear plants were destined to become a main source of electricity and that spent fuel would be reprocessed to recover the fissionable uranium and plutonium they contain. It was then thought that breeding fissionable plutonium from uranium-238 would be necessary to meet the demand for nuclear fuel. With this idea in mind, a prototype pilot plant was designed, incorporating the lessons of Fermi I, to breed uranium into plutonium in the quantities needed by a greatly expanded nuclear electric power industry.

The breeding ratio at the Clinch River breeder, to be constructed near Oak Ridge in Tennessee, was expected to be 1.24, so that the reactor would produce 1.24 atoms of plutonium-239 for every atom of uranium-235 it consumed. The reactor's doubling time, the time required to double the original amount of its nuclear fuel, would have been about 18 years. In the Clinch River breeder, cooling was to be done using liquid sodium, because liquid sodium transfers heat extremely well and can be heated to a high temperature without producing a high pressure. In such a design, the molten sodium emerges from the reactor and flows through an intermediate heat exchanger within the containment building and then is then pumped back through the reactor vessel. This intermediate heat exchanger transfers heat from the first loop of liquid sodium to a second loop of liquid sodium that then flows to the steam generator located outside the containment building. This second stream of liquid sodium provides the heat to the steam generator, which in turn produces high-pressure and high-temperature steam. The steam flows through a conventional turbine and drives the electric generator. The cooled

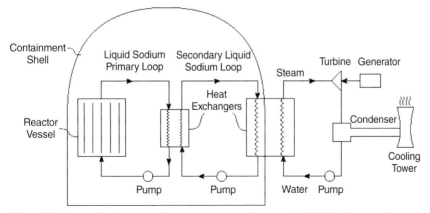

Figure 13.1 Schematic drawing of the liquid sodium and the water and steam flow paths used in a liquid-metal fast breeder reactor. By using two sodium loops, all the radioactive material associated with the reactor, including the primary sodium loop, is completely within the containment shell. Only the secondary liquid sodium loop extends beyond the shell, to boil the water and, via another heat exchanger outside the shell, produce the steam that drives the power-generating turbine.

liquid sodium is pumped back through the intermediate heat exchanger. In this way, electric power and extra fuel are generated at the same time. The Clinch River nuclear reactor was designed to produce 375 MW of net electric power. Figure 13.1 is a schematic drawing of the flow paths for liquid sodium and water in a liquid-metal fast breeder power reactor.

The Clinch River Breeder Reactor was engineered and designed, the main components were built, and the building site was prepared for construction. By the 1980s, though, the tide of support for nuclear power had turned, and the project was canceled in 1983. So were dozens of nuclear power plants that had been planned for many locations in the United States. Instead, most new electric power plants were built to operate with natural gas, which produces carbon dioxide, but less so than coal-fired plants of the same size. Similar non-nuclear approaches to power generation were adopted by many other countries. France, in contrast, was not deterred from building nuclear power plants, and French nuclear power plants now produce over 70% of that country's electric power.

Thermal Breeder Reactors

To breed thorium into fissionable uranium, the high-energy neutrons used by fission in a fastbreeder reactor don't work very well. Instead, neutrons that have been slowed down by repeated impacts with the hydrogen in water or other light atoms are needed. In fact, breeding thorium works best if the neutrons are slowed down so much that they only move around with the velocity they are given by their temperature and their interaction with water molecules. Neutrons moderated this much are called thermalized neutrons, and the resulting breeder reactor is called a thermal breeder. It is basically a normal power reactor whose core is surrounded by a blanket or jacket of thorium. The trouble is that such a light water–moderated breeder reactor has a breeding ratio of less than 1, which means that although it does breed nuclear fuel, it doesn't generate as much new fuel as is used up. To achieve a breeding ratio greater than 1 requires a coolant other than water so that the breeder reactor can be operated at high fission rates and neutron flux levels. Molten salts or helium rather than water could be used as the coolant in such thermal breeder reactors.

Carnot efficiency considerations (see Chapter 9) show that reactors operating at high temperatures yield higher efficiencies for converting heat into electricity than normal, water-cooled reactors do. High-temperature gas–cooled reactors can be built using graphite as the moderating material and helium for cooling, rather than water or molten salts, and such a gas-cooled reactor operates at a much higher temperature than water-cooled reactors can. Helium is especially good for cooling, as it can transport heat faster than any other gas except hydrogen; but, unlike hydrogen, helium cannot burn or explode. The airship *Hindenburg* used hydrogen instead of helium because large supplies of helium were not available to Germany. In the end the *Hindenburg* burned up completely, producing dramatic pictures of exploding hydrogen during an attempted landing in 1937 at Lakehurst, New Jersey. Helium-cooled reactors can run at higher fission rates and consequently have more neutrons and a higher breeding ratio for converting thorium into fissionable uranium than do water-cooled reactors.

A jacket of thorium around the core of a reactor turns it into a breeder reactor. But breeding thorium-232 into uranium-233, which is fissionable, provides another source of fuel for nuclear reactors. Better still, power generation can be carried out at the same time that thorium is being turned into uranium.

Designing and building a new class of power-generating reactors that produce more fuel than they consume will require a lot of engineering and money. So far, the massive investments necessary have not been considered cost-effective. It is currently more economical simply to mine more uranium. But the energy resource that the Earth's thorium represents is enormous and will someday be needed if humanity's demand for energy continues to climb without limit.

Breeder Technology Today and Tomorrow

At the present time, producing uranium from uranium ore and enriching it to 3% uranium-235 is economically more advantageous than breeding uranium into plutonium-239 using uranium-235 enriched to 15%. Consequently, breeder reactor development has been indefinitely postponed. The steady worldwide growth of nuclear power will inevitably increase the demand for uranium, whose price will therefore increase. Increasing the availability of nuclear fuel by a factor of 100 is so attractive that eventually breeder technology will be of great interest again. Predicting when this will occur is difficult, in part because it is a complicated technology that needs to be supported by technologically gifted societies.

In addition, all nuclear technology inevitably means the development of nuclear waste, whose safe storage for millennia remains a conundrum, although a technically solvable one. There is one possibility for breeding nuclear fuel that does not involve the generation of as much waste as either thermal or fast breeder reactors do, and that method involves, surprisingly, nuclear fusion, as will be discussed in Chapter 14. Whether produced by fission or fusion processes, the long-term storage of nuclear waste may be a matter that is more difficult to solve politically than technically. Other technologies, especially solar, are actively being advanced now, and such renewable energy sources are the preferred development path being pursued by many nations. In the end, it is likely that all energy sources, including breeder reactors, are going to be needed, but for the next few decades, breeder technology will probably have a low priority. But breeding offers a prospect that is unlikely to be ignored indefinitely.

Recommended Reading

Judd, A.M. *Fast Breeder Reactors: An Engineering Introduction.* New York: Pergamon, 1981.

Sharefkin, Mark F. *The Fast Breeder Reactor Decision: an analysis of limits and the limits of analysis: a study.* Washington, DC: U.S. GPO, 1976

Shultis, J. Kenneth. *Fundamentals of Nuclear Science and Engineering.* New York: Marcel Dekker, 2002.

14
Nuclear Fusion: Engine of the Sun

As for the future, we may say, with equal certainty, that
inhabitants of the earth cannot continue to enjoy the light and
heat essential to their life, for many million years longer, unless
sources now unknown to us are prepared in the great storehouse
of creation.

Sir William Thompson (Lord Kelvin), *Age of the Sun's Heat*, 1862

The "great storehouse of creation" Lord Kelvin mentions does contain an energy surprise: nuclear fusion. Kelvin was among the foremost scientists of his time, but nuclear energy was still unknown when he was trying to reconcile the geological age of the Earth and the source of the energy that the sun so copiously supplies. Kelvin's calculations showed that even if made entirely of coal and oxygen, chemical energy could not keep the sun going for more than a few thousand years. He also found that neither gravitational collapse nor swarms of incoming comets could have powered the sun for the eons that geology proved it must have been shining.

We have already seen that nuclear fission of uranium produces greater than three million times more energy per pound or kilogram than burning coal does. Beyond nuclear fission lies the potential of nuclear fusion. In nuclear fusion, light atoms such as hydrogen are merged together, forming heavier atoms. This process produces energy the same way that breaking apart heavy uranium atoms does. Pound for pound, fusing hydrogen to make helium produces two orders of magnitude more energy than does fissioning uranium to make barium and krypton, and it makes 300 million times more than does burning carbon to make carbon dioxide. In addition, fusion reactors don't aggravate the nuclear weapons proliferation problem the way fission reactors do. That's the energy surprise in the great storehouse of creation. The question is: How can this energy surprise be harnessed on Earth?

Cold Fusion versus Cool Fusion versus Hot Fusion

One apocryphal method, cold fusion, was loudly trumpeted because it involved neither high temperatures nor high voltages. This fusion process has never

Energy Demand and Climate Change: Issues and Resolutions. Franklin Hadley Cocks
© 2009 WILEY-VCH Verlag GmbH & Co. KGaA, Weinheim
ISBN: 978-3-527-32446-0

been proven, although reports of anomalous heat generation (possibly from hydrogen embrittlement and subsequent cracking of the palladium rods used in this process) have lingered for years. Although the cold fusion concept has not panned out, it did stimulate renewed interest in muon-catalyzed fusion, which is a proven way to produce fusion at or near room temperature. Muon-catalyzed fusion is sometimes called cool fusion. Muons are decay products of the pions that enable protons to stick together in atomic nuclei, as discussed in Chapter 5. Muons can be thought of as heavy electrons, because they have the same negative electrical charge as electrons but weigh more than 200 times as much. Unlike electrons, which last indefinitely, free muons last only 2.2 microseconds. If, in this short time, a muon is approached simultaneously by two nuclei of hydrogen, one a deuterium nucleus (D), with one proton and one neutron, and one a tritium nucleus (T), with one proton and two neutrons, these two nuclei will fuse together. The muon is left unscathed, thanks to the magic of subatomic physics.

The problem is that muons don't last very long and take a lot of energy to produce. Muons are created by cosmic rays interacting with upper-air molecules, but there are nowhere near enough natural muons to produce any significant D-T fusion. Producing muons in quantity requires much more energy than they can generate by inducing D-T fusion. And the D-T-muon *ménage a trois* is hard to arrange. Muons mostly end up simply decaying to form various particles, including electrons and various kinds of neutrinos, those ghostly particles that the sun produces and that pass through matter so easily that it takes a light-year-thick layer of lead to stop them.

Instead of muon-catalyzed fusion, brute force D-T fusion is currently the method of choice for making hydrogen bombs. Hydrogen bombs can be much more powerful than uranium bombs, but they need atomic bomb explosions to set them off. Making D-T fusion happen controllably on a large scale is even more difficult than making hydrogen bombs, but the driving force for trying to do so is the goal of harnessing, for peaceful purposes, the potent source of nuclear energy that fusion represents. On the upside, nuclear energy, whether from fission or fusion reactors, generates electric power without producing carbon dioxide and its global-warming side effects. On the downside, nuclear fission reactors are attractive targets for terrorists, and they generate radioactive waste that must be stored for millennia. D-T nuclear fusion also produces radioactive refuse, but not as much as fission does. In addition, a fusion reactor, unlike a fission reactor, isn't very useful for making atomic bombs, so it doesn't aggravate the weapons proliferation problem, either.

Even with the most achievable fusion reaction, that is, fusing deuterium (D) with tritium (T) to form helium, there are problems. Neutron radiation is one of them. D-T fusion produces normal helium, whose nucleus has two protons and two neutrons, plus a free, energetic neutron moving at a speed not far below that of light. These energetic neutrons can make other objects radioactive as well as causing damage to whatever they hit. As it happens, D-T fusion is not the main fusion reaction that goes on in the sun. The primary fusion reaction of the sun involves simple hydrogen nuclei, and its primary end product is also normal helium, but the sequence of fusion steps needed is convoluted. Fusing simple hydrogen on Earth to produce helium is a lot more difficult than fusing deuterium and tritium. Nobody today seriously suggests that fusing simple hydrogen is going to solve our energy problem anytime soon, but fusing deuterium and tritium just might do the trick.

Making Fusion Happen

The process of fusing atoms together is radically different than that of breaking heavy atoms apart. But even after more than 50 years of effort, nobody has been able to get significantly more energy out of fusion than they put into it, except by making a hydrogen bomb. Unless a way is discovered to produce lots of extra energy *controllably*, thermonuclear fusion will be useful only for weapons of war and mass destruction. Producing controllable nuclear fusion on a grand scale without a nuclear explosion – that's the problem.

The energy from the fusion of deuterium and tritium is commonly thought to offer unlimited energy potential. About one out of every 6700 hydrogen atoms in seawater are deuterium. But tritium is rare on our planet. Heavy water, which can be made in nuclear reactors, consists of deuterium and oxygen. Deuterium is stable, not radioactive, relatively common, and heavier than normal hydrogen. It is denser than the water we drink water because normal hydrogen (H) contains only a single proton and no neutrons, unlike deuterium with its nucleus of a proton plus a neutron. Water that contains lots of deuterium is termed heavy water because it is denser than normal water. Tritium, with one proton and two neutrons, is radioactive and rare. Tritium decays into helium-3 with a half-life of 12.32 years. This means that the tritium in any application that uses it, such as hydrogen bombs, steadily turns itself into inert helium-3.

D-T fusion is by far the easiest fusion reaction to achieve. Keeping this reaction supplied with deuterium is not difficult. Deuterium is obtained from water, usually by electrolysis. A lot of electricity is needed, which is why the Nazis produced deuterium in Norway, with its numerous hydroelectric power plants. Using heavy water, it is possible to operate a reactor that does not need enriched fuel. CANDU (CANadian-Deuterium-Uranium) reactors were developed in Canada and operate with heavy water and natural, unenriched uranium.

Keeping the D-T reaction supplied with tritium, however, is a problem. Tritium can be produced by the interaction of the neutrons created by D-T fusion reacting with lithium, especially the isotope lithium-6 (three protons with three neutrons). Lithium-6 comprises 7.42% of all lithium, and obtaining more tritium than is used up by the D-T fusion process requires natural lithium enriched with lithium-6. Tritium generation uses up lithium, but the amount of available lithium on the Earth is sufficient to enable D-T power plants to produce far more energy than all fossil fuels put together. Theoretically, in a fusion power plant, a jacket of liquid lithium around the fusion chamber is heated by the absorption of the energetic neutrons from fusing deuterium and tritium. The resulting molten lithium supplies the heat needed to boil water and generate electricity. One engineering problem: molten lithium is extremely flammable, and even solid lithium can burst into flame on contact with water. A burning fusion power plant would be almost as nasty as a burning nuclear fission power plant. The lithium fire could not be put out with water, and all the tritium it contained would be released into the environment. On the sun, nuclear fusion reactions are driven by the combination of high temperature and pressure. The sun is a ball of plasma, meaning that electrons have been stripped from atoms, and these stripped atoms, called ions, and the resulting free electrons are held together by the sun's gravity. In the core of the sun, vast amounts of hydrogen are fused via a complicated sequence of nuclear steps and produce prodigious amounts of energy, which then percolates up to the surface and is radiated into space. Fusing light atoms on Earth using an enormous volume of plasma held together by gravity is not possible. The question then becomes, can D-T fusion be done controllably on a large scale on Earth? For more than 50 years serious attempts have been made to harness the D-T fusion reaction. The standing joke in the fusion community is that success is always just 20 years in the future. The truth appears to be that on a large scale this feat is monumentally difficult, perhaps beyond the scientific or economic ability of any single nation. The world's largest thermonuclear research project is ITER (International Thermonuclear Energy Reaetor). It is jointly sponsored by several nations acting in concert, but the ITER machine is being built in France. One of its goals is simply to break even and produce at least as much energy as it consumes. How is it supposed to work?

ITER, Tokamaks, Magnetic Fields, and Fusion

There are three different engineering approaches to making the nuclei of a light element such as hydrogen fuse together: magnetic compression, inertial confinement, and accelerator fusion. Magnetic compression creates dense plasma at extremely high temperatures. In this plasma the energy of the deuterium and tritium ions is high enough to fuse them when they hit each other. The trick is to get the plasma density and the temperature high enough to produce energy by fusion faster than it is lost by radiation from the hot plasma. The ignition temperature is that temperature at which the heat produced by fusion is greater than the heat lost from the plasma. For D-T fusion the ignition temperature is about 70 million degrees Kelvin! Incredibly, this fantastically high temperature is lower than for any other fusion reaction. In D-D or T-T fusion, for example, the ignition temperature is around 200 million degrees Kelvin. At either of these temperatures, nothing is solid except exotic nuclear substances such as the neutron-degenerate matter postulated to exist in neutron stars. D-T is the fusion process being attempted by the ITER project.

The walls of the fusion chamber can't stand temperatures even $1/10,000^{th}$ as high as 70 million degrees. One solution to the problem of losing energy to the walls of the fusion chamber makes use of the fact that ions moving in a magnetic field tend to spiral around the magnetic field lines. If the magnetic field is strong enough, the plasma doesn't touch the walls of the chamber. Because the rate of fusion goes up as the plasma becomes denser, fusion is achieved if magnetic fields are also used to compress the plasma as well as contain it. Many approaches have been tried for this purpose. Initial experiments used cylindrical or tubular magnet designs, but these were found to lose too much energy from the ends of the tubes. The simplest answer was to close the tube upon itself so that there are no ends. The most famous version is the so-called *tokamak* design, after the Russian words for "magnetic donut." The ITER project, for example, will use a tokamak configuration.

In tokamak reactors like that of ITER, the fusion chamber is similar to a hollow torus or donut. Inside this hollow donut, the plasma is contained and compressed by magnetic fields produced by wire coils spiraling around the outside of the donut, and there are no ends from which the plasma can escape. In addition, current is induced in the plasma itself by another set of wires arranged so that the donut-shaped plasma acts like part of a transformer. The current induced in the plasma by the transformer wire coils heats and compresses the plasma. The tokamak approach does indeed produce D-T fusion, but the cost in energy (and money!!) is very high. The scale of the effort going into the ITER project may be imagined from the sketch of the tokamak core being constructed as part of this project, as shown in Figure 14.1.

Tokamak power plants are still a long way off, seemingly always the proverbial 20 years or more. This situation may be improving, and it appears, at least in some experiments, that more energy may be produced, momentarily, than is consumed. These experiments are nowhere nearly successful enough for tokamaks to be used as power plants anytime soon. To make a working fusion power plant, a liquid blanket of lithium, about a meter thick, surrounds the torus fusion chamber, absorbing energy from the energetic neutrons produced by the fusion reaction and also producing more tritium. To reach the lithium, the neutrons must first pass through the outer shell of the chamber, which should not itself absorb them. Zirconium-niobium alloys appear to be possible metals for this purpose because they are compatible with molten lithium and don't strongly absorb high-energy neutrons. Of course, the radiation damage to the niobium is extreme, and its mechanical properties will be substantially altered by the neutrons passing through it. It is known, for example, that when neutrons bombard ultra-pure niobium, its mechanical properties degrade, even with bombardment levels thousands of times lower than those experienced by the walls of a tokamak power plant. Niobium is also expensive, and experimental units like the one at the ITER facility do not have a molten lithium blanket surrounding it. The walls of the ITER machine are made from stainless steel, which is much cheaper than niobium. It turns out that fusion reactors can also function as breeder reactors, so let's look briefly at this possibility next.

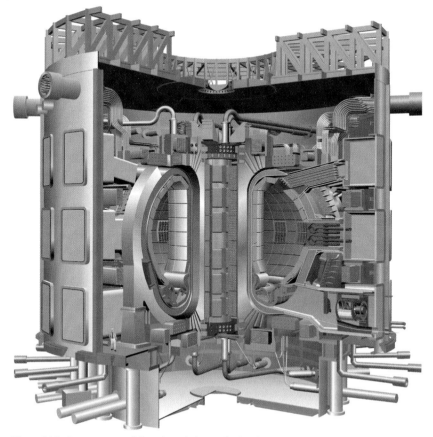

Figure 14.1 Cross-section of the tokamak fusion facility for
the ITER project. The scale of this facility can be seen by
noting the size of the person standing in front.

The Combined Fusion–Breeding–Fission Process

As we discussed in Chapter 13, fast neutrons can be used to convert unfissionable
uranium-238 into fissionable plutonium-239. When a fusion reactor is working, the
fusion of deuterium and tritium produces lots of fast neutrons and helium atoms.
If these fast neutrons are absorbed by a thick layer of uranium-238 on a wall of
the fusion chamber, they convert some of this uranium into fissionable plutonium-
239. Better still, the heat associated with this process can diffuse through a second,
much thinner wall that can be made of stainless steel rather than zirconium-
niobium. This heat can then be used to boil water and generate electricity in the
usual way. In addition, the neutron flux from fusing D-T can also cause fission
of the plutonium that is being produced in the thick layer of uranium-238, and
this fissioning plutonium adds still more heat. In this way, a fusion reactor can

be used to generate electricity as well as fuel for fission reactors. Naturally, the resulting combined fusion–breeding process is complex, and engineering it properly would be challenging indeed. However, it is a process that, at least potentially, offers enormous benefits. By absorbing fast neutrons in uranium-238, for example, the problems of radiation damage and mechanical property changes in the outer wall are avoided.

One shortcoming of this combined fusion/breeding process is that it does not produce tritium, which would therefore have to be produced by a separate nuclear reaction. The key development needed, of course, is a viable fusion process. Fortunately, magnetic containment fusion is not the only fusion possibility. The concept of using the fast neutrons produced via nuclear fusion to transmute uranium-238 to plutonium-239 is applicable not only to magnetic confinement fusion but also to inertial confinement fusion, which is described next.

Inertial Confinement Fusion

A second method of producing fusion is so-called inertial confinement. This means hitting the D and T atoms with so much energy so fast that they don't have time to fly away before fusion starts. This is the approach used in hydrogen bombs, where the high blast of energy from an exploding uranium or plutonium bomb acts as an igniter for the fusion reaction. The radiation from the nuclear fission detonation causes the D-T to fuse before the blast wave from the uranium or plutonium bomb blows everything to smithereens. Such an approach is obviously not practical for a fusion power plant, so instead of using nuclear bombs, massed banks of high-power lasers fire simultaneously at small "pellets" containing tritium and deuterium. One problem with this approach is finding a material to contain the gaseous deuterium and tritium, because the atoms in this container get blasted, too, and soak up energy without contributing to the fusion reaction. Freezing the deuterium and tritium into a solid pellet doesn't work well because it can vaporize into gas before the laser beam can hit it. Fusion can be produced by the focused laser approach, but once again more energy goes in than comes out. Also, this process must be repeated every few seconds to produce a large, steady supply of energy.

Accelerator Fusion

A third method, which has not been investigated as thoroughly as magnetic and inertial confinement methods, involves using electric fields to accelerate the electrically charged D and T ions to high velocities. The advantage of this method is that the temperature equivalent of voltage is very high. It takes less than 10,000 volts to cause D-T fusion to occur, not a temperature of 70 million degrees. The disadvantage of this accelerator method is the difficulty of getting large numbers of deuterium and tritium ions to collide with each other. One suggested approach to solving this problem is the concept of shooting a beam of deuterium ions at a frozen, solid wall of tritium, or vice versa. One big problem with this approach is that hydrogen, whether deuterium, tritium, or regular hydrogen, is solid only at extremely low temperatures (at normal pressures, anyway). An incident ion beam would rapidly vaporize the solid, frozen wall. Of course, solid, high-melting-point compounds of hydrogen isotopes could be used instead of frozen hydrogen isotopes, but even so the incident power of the ion beam needed to make a power plant work is stupendous. Unless spread out, this ion beam would melt and vaporize any known solid substance. Spreading the ion beam out over a wide area certainly helps, but for a 1000-MW power plant, the beam would have to be dispersed over a wall about the size of a football field to avoid vaporizing the surface the spread-out ion beam impacted.

Yet another approach to ion-beam fusion is to send two beams, one of deuterium and one of tritium, directly at each other. The problem here is that at the maximum beam densities producible by known methods, most of the ions miss each other. Another approach is to oscillate the ions back and forth inside a spherical electrostatic field so that each ion gets many chances to fuse with another ion. In the end, ion-beam fusion has never been made to work on a large scale. But producing fusion using electric fields to accelerate deuterium and tritium ions, for all its problems, has the advantage that the voltage needed to produce fusion is relatively low. Accelerator fusion permits consideration of fusion reactions that would be virtually impossible to achieve by the magnetic confinement method being planned on a grand scale for the ITER facility. In particular, accelerator fusion allows the fusion of deuterium with helium-3 to be considered, and this fusion reaction offers some very special advantages.

Fusion of Helium-3 and Deuterium

The ignition temperature for the D-T fusion reaction is about 70 million degrees Kelvin, while the ignition temperature for the plasma fusion of deuterium and deuterium is about 200 million degrees. The ignition temperature for fusing deuterium with helium-3 is higher still, at about 700 million degrees. But in terms of voltage, just over 100,000 volts are needed to make ions move at velocities equivalent to 700 million degrees, and 100,000 volts can be easily produced with existing technology. D-He^3 fusion has the advantage that it produces energetic protons, not the neutrons that both D-T and D-D fusion produce. Protons scarcely produce radioactive isotopes at all in comparison to neutrons. Neutrons interact with atoms easily, generating lots of isotopes in the process, and many of these isotopes are radioactive. When deuterium is fused with helium-3, energy is released in the form of high-energy protons and helium-4 ions, not neutrons. Protons are just the atomic nuclei of regular hydrogen. Protons and helium ions, being electrically charged, do not enter other nuclei easily and thus do not readily produce radioactive isotopes. A high-energy proton can occasionally knock loose a neutron if it strikes a nucleus head-on, but this so-called *spallation* reaction does not happen often. Typically, it takes more than 1000 high-energy protons to produce one neutron via spallation, so protons don't make things radioactive nearly as easily as neutrons do.

He^3-D fusion can theoretically occur via either high voltages or very high temperatures, but thermally driven, high-temperature He^3-D fusion also produces some D-D fusion, and this side reaction releases neutrons. Voltage-driven fusion is the way to go, since the temperatures required for He^3-D fusion are so high. No matter how the fusion of helium-3 and deuterium is carried out, the high-energy protons and helium-4 ions it produces are charged particles moving at high speed. Magnetic fields can be used to direct them into forming a beam of current at very high voltage. A uniform magnetic field applied perpendicularly to this beam will separate the electrons from the high-energy protons and helium-4 ions. Because helium-4 ions and protons have electric charges opposite to that of electrons, their paths will be curved in opposite directions by applied magnetic field and can be directed to different collection points from that of electrons. This charge separation gives rise to a high-voltage electric current, and no thermodynamic cycle is required to produce this electric power. This fact means that the efficiency of electric power generation in this way could be very high, since the Carnot efficiency limit does not apply, as it does to steam power plants. In contrast to D-T or D-D fusion reactions that end up using boiling water to produce electricity, He^3-D fusion energy could be converted into electric power at efficiencies that, at least theoretically, could approach 100%.

One trouble with the D-He^3 concept is that there is not much helium-3 on the Earth. But there is a lot of it on the moon.

Lunar Resources of Helium-3

In the 1980s, lunar helium-3 was suggested as a resource for fusion power. For more than four billion years the solar wind has been blowing against the surface of the moon, depositing helium-3 (and helium-4) onto its surface. On our world, these fast-moving ions from the sun, which also include lots of protons, never hit the ground. Instead, they spiral down the Earth's magnetic field until they strike the upper atmosphere, creating glowing curtains of auroral light that shimmer and ripple in the dark polar sky before dispersing into the air and eventually being lost to space because they are so light in weight. On the moon, this helium is buried in surface lunar soil.

Samples collected during the *Apollo 11* and *Apollo 12* missions showed that moon dust contains helium. Most of it is helium-4, but 0.04% is helium-3. The physical properties of helium-3 and helium-4 are quite different. The boiling point of liquid helium-3 under a pressure of one atmosphere is 3.19 K while that of helium-4 is 4.21 K, so separation by differential boiling is easy.

The moon's surface near the equator contains about 20 parts per billion of helium-3, decreasing toward the poles, but the possibility that the shadowed polar regions might have especially rich helium-3 deposits has not been recognized before. There may be 10 times, possibly even 100 times, higher levels of helium-3 in the permanently shadowed polar craters than on the sunlit surface. Even though sunlight doesn't shine into deep polar lunar craters, the solar wind is ionized and swirls every which way due to the ever-shifting magnetic fields in space. Helium ions can reach the permanently shadowed craters of the lunar poles even though the sun never shines there. On the sunlit surface, the temperature can exceed 125 °C and can "boil" away 99% of the trapped helium. In cryogenically cold polar craters, almost all of it stays put.

Even if the perpetually cold floors of permanently shadowed craters received only 10% as much solar wind as the sunlit lunar surface, they could still contain 10 times as much helium-3 per gram of lunar soil because they keep virtually all they get. Both lunar poles have permanently shadowed craters, but the south lunar pole has the most favorable ones.

Undoubtedly there are numerous scientifically important things to be found on the moon, but helium-3 is one of the very few lunar resources economically valuable enough to be worth carrying back to Earth, and it is possibly the only one. It has been calculated that the energy it would take to go to the moon, recover helium-3, and transport it back to Earth is less than 1% of the energy this He^3 could produce by fusion with D. In comparison, mining and transporting coal uses up about 6% of the energy it produces when burned. A lunar base will be required for harvesting helium-3. Any base on the moon will need every natural resource it can find, including helium-3, subsurface frost, and soil, to fabricate structures and extract oxygen.

A permanent base has good use for sunlight. Either electric power from solar cells or nuclear power or both would be needed for a long-lasting lunar base. At most locations the two-week-long lunar night presents a real problem. At first glance, polar craters might seem to be the worst possible place for solar power. However, the moon is not tilted as the Earth is, and the outer wall of the Shackleton crater, located almost exactly at the south pole, is in sunlight 80% of the time. The surface of the rest of the moon is mostly illuminated just 50% of the time. That may be why the recently proposed NASA lunar base is to be located at the outer wall of the Shackleton crater. The cost of building a base anywhere on the moon will be huge, but if we can contrive a way of harnessing the He^3-D fusion reaction, the payoff would be enormous. Meanwhile, a robotic mission to measure the helium-3 and water stored in permanently shadowed lunar polar craters could be carried out at moderate cost. Of course, sending a robot explorer into the dark shadows of a crater as cold as −200 °C presents an engineering challenge. But surely no more so than did the *Apollo* or Mars missions. An on-the-surface survey of the He^3 and frost hiding in the stygian darkness at the southern lunar pole could be well worth its price.

Producing energy on a large scale by fusing light isotopes such as deuterium and tritium or deuterium and helium-3 would be a monumental achievement for humanity. Whether or not this can be accomplished remains to be seen. Right now, it is still very much an energy dream, but it is one with great promise, even if it always seems to be 20 years in the future.

Recommended Reading

Bishop, Amasa S. *Project Sherwood: The U.S. Program in Controlled Fusion*, Reading, MA: Addison-Wesley, 1958. (This very early book reviews the whole gamut of possible methods for controlled thermonuclear fusion.)

Clark, R.E.H. and D. Reiter. *Nuclear Fusion Research: Understanding Plasma-Surface Interactions*. Berlin: Springer, 2005.

Ford, Kenneth W. *The World of Elementary Particles*. New York: Blaisdell, 1963.

Woods, L.C. *Theory of Tokamak Transport: New Aspects for Nuclear Fusion Reactor Design*. Weinheim: Wiley-VCH, 2006.

See also the Web site for the ITER project: http://www.iter.org/a/index_use_5.htm

15
Power from the Ocean: Thermal and Salinity Gradients

*This invention relates to method and apparatus whereby we
have succeeded in obtaining power from the difference in
temperature between the surface water of tropical seas and the
water at great depths.*

Georges Claude and Paul Boucherot, U.S. Patent Number
2,006,985, 2 July 1935

The oceans hold immense amounts of energy, but the problem is finding a
way to use this energy to generate electricity. One way involves water at different temperatures. In 1881 the French physician and physicist Jacques-Arsène
D'Arsonval was the first to suggest using temperature gradients in the ocean
to generate electricity. He pointed out that the differences in temperature
between surface equatorial ocean water and deep ocean water could be used
to generate electric power. Speculating on the various natural sources of temperature gradients that exist on earth, such as those involving volcanoes and
glaciers, he feared "encroaching upon the territory of M. Jules Verne." The use
of gradients in salt concentration to generate electricity is a more recent conceptual development. Heat flows from hot regions to cold regions, and dissolved salts move from high concentrations to dilute ones. In both cases some
of the energy represented by differences in temperature or salt concentration
can be converted into electricity. Let's consider first the possibility of generating electric power from ocean thermal gradients and see how that process
works.

Electric Power from Ocean Thermal Gradients

Warm ocean water has always been more plentiful by far than hot geothermal
water, but attempts to tap ocean energy resources to generate electric power have
run into many problems. It was not until 1929 that D'Arsonval's student, the
French inventor and entrepreneur Georges Claude, built the world's first ocean
thermal power plant at Matanzas Bay on the north coast of Cuba, less than a two-
hour drive east of Havana.

Energy Demand and Climate Change: Issues and Resolutions. Franklin Hadley Cocks
© 2009 WILEY-VCH Verlag GmbH & Co. KGaA, Weinheim
ISBN: 978-3-527-32446-0

Claude was a determined man who made his fortune by co-founding a company to liquefy air and produce pure oxygen. Using the rare gases that were byproducts from his air liquefaction process, he increased his wealth by developing a new type of advertising sign. These new signs used the bright orange glow that occurs when an electric discharge is sent through the rare gas neon at low pressure (about 1% of normal atmospheric pressure). Claude was the inventor of neon signs.

The ocean thermal energy plant Claude constructed on the shores of Matanzas Bay encountered a host of difficulties. He persisted nonetheless and finally succeeded in producing enough electricity to power the equivalent of 220 light bulbs, each of 100 watts, by using the difference in the vapor pressure between warm surface water and cool deep water to run a generator. And he accomplished this feat using a temperature differential of only about 14 °C, which gave him a pressure difference of only about one quarter of a pound per square inch! Unfortunately, the electric power needed to pump ocean water to Claude's plant was greater than the electricity generated, but he did at least prove the concept of using ocean thermal energy conversion (OTEC) to produce electricity. How exactly did Georges Claude's power plant work?

Warm surface ocean water can reach at least 81 °F (27 °C), while water 1 km down can be as cold as 39 °F (4 °C). The vapor pressure of water increases exponentially with temperature, but the difference in vapor pressure between water at 27 °C and 4 °C is still only about 0.4 pounds per square inch (2758 pascals). The difference in vapor pressure between water in the ocean of different temperatures is very small compared to that in steam plants, which is usually around 2500 pounds per square inch (over 17 million pascals), but the operating principle is the same for ocean thermal plants as it is for normal power plants. The vapor pressure difference of water at two different temperatures can be made to spin a turbine connected to an electric generator just as high-pressure steam can. However, don't forget Carnot (see Chapter 9): 27 °C is about 300 °K and 4 °C is about 277 °K, so the maximum possible efficiency of a "steam" or any heat engine operating between these two temperatures is $(1 - 277 / 300) = 7.7\%$. This low maximum efficiency means that a very large volume of water vapor is needed to produce a modest amount of power.

The design for a power plant needed to produce electricity using the difference in vapor pressure between warm and cold seawater is shown in Figure 15.1. The caption for this figure sets out in detail the process by which electric power is generated. Basically, the passage through a turbine of the higher pressure of water vapor above the warm water is the driving force that keeps an electric generator spinning.

Unfortunately, Claude's power plant soon failed, partly due to the repeated breakage of the very large steel piping he used to bring cold ocean water to the shores of Matanzas Bay from about 2000 feet (600 meters) down and 6000 feet (1850 meters) away. Finally abandoning this shore-based facility and its associated giant water pipe, he built a second plant onboard a ship and sailed to just off the coast of Brazil. At this new location the length of piping necessary to bring cold water from the ocean depths was a lot less than at Matanzas Bay, and he could

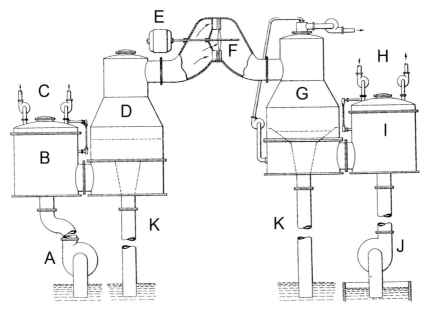

Figure 15.1 Design of the warm–cold seawater power plant of Georges Claude. The warm-water circuit is on the left, and the cold-water circuit is on the right. The warm water is supplied to chamber B by pump A and degassed by vacuum pumps C. In this warm-water circuit, the seawater then enters evaporator chamber D, where it "boils" under the low pressure there (less than 10% of atmospheric pressure). The resulting "steam" then passes through turbine F, which drives electric generator E. This vapor is then condensed in chamber G by cold water from chamber I. Chamber I is supplied with cold ocean water by pump J, which is also used to remove dissolved gas from the cold seawater using vacuum pumps H. If the dissolved gas in the cold water were not removed, the pressure differential across turbine F would be impaired. Excess water from both the warm- and cold-water circuits is returned to the ocean through drain pipes K. Because rapid evaporation lowers the warm water temperature, a large warm water flow rate is needed to minimize this temperature drop. Adapted from Claude and Boucherot, U.S. patent number 2,006,985.

also take advantage of the greater temperature differences between the surface and deep ocean waters there. Always the entrepreneur, he intended to use the electricity he produced to manufacture ice, but this shipboard power and ice plant also suffered from numerous problems, especially corrosion. In a grand finale of frustration, Claude apparently sank his ocean-going power plant into the cold, dark ocean depths. His attempt to generate electricity using ocean thermal energy was over for good. At the end of World War II, he was convicted of being a Nazi collaborator, stripped of his fortune and honors, and sentenced to life in prison. After serving three years, he was released at the petition of the French Academy of Sciences. He died, unheralded, in Paris in 1960, but his dream of harvesting power from temperature differences in ocean water lives on.

The enormous potential for generating electric power from hot versus cold seawater has kept Claude's dream alive. Chlorofluorocarbon gases (and more recently hydrochlorofluorocarbon gases), like those used in air-conditioning systems, as well as ammonia (NH_3) have been proposed as working media for ocean thermal energy systems rather than water vapor, because they can provide much higher vapor pressures than water does. Using these gases, heat exchangers must be interposed to heat and cool the ammonia or the chlorofluorocarbon gases using ocean water. Operating in a closed system, the ammonia or chlorofluorocarbon does not escape. In so-called open systems like that used by Claude, the water is not recycled. In closed systems, the chlorofluorocarbon or ammonia is recycled. Claude had less than one pound per square inch of pressure difference to work with. For the temperature differences that he had available, ammonia produces a pressure difference more than 150 times greater than ocean water does. Chlorofluorocarbons are similar. Remember, Carnot's maximum efficiency rule still applies no matter what working fluid is used, whether water vapor, ammonia, chlorofluorocarbons, or anything else. A power plant using ammonia or chlorofluorocarbons will have the same maximum efficiency limit as one that uses water vapor, but the turbines needed to spin the electric generators and the associated piping for the working vapor can be a lot smaller than they would need to be using water vapor because of the higher vapor pressure.

To decrease the length of the needed pipes and the associated pumping energy required to bring cold water to the surface, it is possible to consider floating ocean thermal power plants, just as Claude did. At the edge of the Gulf Stream, for example, cold water and warm water are close together. But the Gulf Stream shifts constantly, and the power plant must follow its meanderings. Also, electricity is generated far from shore. What to do with it? This electricity could electrolyze water to make hydrogen or produce metallic sodium metal from the sodium chloride dissolved in seawater. This hydrogen or sodium metal could then be shipped back to shore. Sodium metal can be used to make hydrogen simply by spraying it with water. All of this adds a lot to cost as well as complexity. Even worse, ocean storms occur from time to time, and such a floating power plant might sink, which is one thing at least that does not threaten a power plant on land.

One concept for generating electricity from thermal gradients in water that does not seem to have been explored very much involves cold ocean water in combination with solar water-heating of fresh water. There might be several advantages to such a scheme compared with using naturally warmed surface ocean water. For one thing, the water temperatures that could be produced using flat panel solar thermal collectors with very high absorptivity–low emissivity coatings (see Chapter 6) combined with glass cover plates can boil water or produce ammonia or chlorofluorocarbon vapor having a very substantial pressure. The vapor pressure of liquid ammonia at $100\,°C$ is more than 60 times that of water, so the turbine size of a low-temperature power plant that uses ammonia as its working gas can be a lot smaller than one that uses water vapor. Another alternative is using warm fresh water versus cool ocean water, which increases the pressure difference that is available, because the vapor pressure of water above ocean water is less than that above fresh water, even if they are both at the same temperature (see the next section on osmosis).

There are numerous seashore locations that enjoy copious sunshine, and many of them are on islands that rely entirely on imported fossil fuel for electric power. It is possible to heat water in solar ponds (see Chapter 6) to much higher temperatures than can be reached by surface ocean water, and as the water temperature goes up, so does power plant efficiency. The question of what to do at night still arises, but unlike solar cells, the energy for a solar-powered thermal plant can be stored in the form of hot water, and this is not difficult, as millions of hot water tanks prove every day. Solar ponds don't have to be near the ocean, but if they are, the availability of cool ocean water would be an advantage in operating them. Undoubtedly, there will be unexpected problems in operating solar-pond power plants, but the concept of solar water heating with cool ocean water seems to be worth further evaluation, especially for use on islands. If cool ocean water is not available, then solar ponds can be designed to have very salty, dense water near the bottom and less-dense, fresh water on the top, as was discussed in Chapter 6. The temperature difference between the surface water and the warmer dense, salty water near the blackened bottom of the solar pond can then be used to generate electricity. By using a separate pond for heating water, this water can be made far warmer than would normally be available from the surface water of the ocean, so that ocean water could be used on the cool-side of an ocean thermal power station, thereby avoiding all the difficulty of bringing cold water up from great depths. By using chlorofluorocarbons or ammonia rather than water vapor as the working fluid, the size of the necessary turbine and associate piping would be dramatically reduced, thereby lowering the total cost. Small prototype thermal pond power stations have been built and they do work, but so far, they are well away from being economically viable compared with burning coal to boil water.

The small volume of a pond or a dammed ocean inlet compared with the ocean itself limits the power-generating ability of such facilities. The water basin for producing the necessary hot water for a one-megawatt power plant would have to cover many acres or hectares of area. Building these plants would be practical only if standard, packaged designs and equipment were mass produced to reduce the cost to a minimum. No such design and development effort is now underway. And it would take 1000 such facilities to reproduce the power from one typical coal-fired power plant. But they would be renewable, non-polluting, and inexhaustible sources of electric power. For supplying power to small islands, a one-megawatt facility could make a big contribution.

The energy resources of the oceans are very large, but so are the difficulties in using them. Unlike the energy that is present in the high-temperature rock deep in the bowels of the Earth, ocean thermal energy is much more accessible. So far, though, no ocean thermal energy conversion (OTEC) plant has proven to be economically viable, but OTEC systems offer notable extra benefits besides electric power. Cold, nutrient-rich deep seawater can be used for aquaculture and for air conditioning. OTEC systems can produce fresh water as the water vapor they produce is condensed. One drawback is that deep-ocean water releases carbon dioxide when brought to the surface and warmed up. In spite of the difficulties and because of its potential, a number of attempts to develop OTEC systems have been made.

For three months in 1979 an OTEC system aboard a Navy barge anchored off Keahole Point, Hawaii, produced a net power output of up to 17,000 watts, not quite as much as Claude's plant on Matanzas Bay did. This project was followed in 1980 by OTEC-1, a facility constructed on the Kona coast of Hawaii for evaluating different OTEC concepts, including using cold, nutrient-rich ocean water for air conditioning and aquaculture. In 1981 a 100,000-watt plant was built by Japanese interests in the Republic of Nauru, a Pacific Ocean nation. Production of OTEC power from a facility on the southern coast of India, in Tuticorin in the Indian State of Tamil Nadu, was announced in 2008.

The total theoretical power-generating potential of ocean thermal energy conversion systems is enormous, but the number of locations where OTEC plants might possibly be practical is limited. However, the engineering of OTEC systems is being improved to the point that electric power in this way might become economically viable, especially for islands, if the will and money are there to make this happen.

In addition to the OTEC concept, there is another renewable energy resource based on gradients in salt concentration rather than temperature. This second gradient power scheme doesn't have the same location problems that OTEC plants do. It involves harnessing the power that can be generated from diluting salt water with fresh water. This salinity gradient method has been far less investigated than ocean thermal energy conversion, so let's consider it next.

Electric Power from Ocean Salinity Gradients

Long ago it was discovered that pig bladders tied shut while still filled with urine and thrown into fresh water would swell up, even to the point of bursting. This swelling happens because fresh water diffuses through the bladder wall and dilutes the salty urine inside. In the 19th century self-dilution phenomena were studied scientifically, and it was discovered that the pressure produced by this dilution process depends on the concentration of the salty solution in the bladder. Other membranes from both plants and animals, such as leaves and skin, were soon shown to exhibit similar behavior. Water molecules and salt molecules can move at different rates through different membrane materials. In 1854 this effect was termed osmosis. Nowadays, synthetic plastics rather than natural membranes are used in devices such as kidney dialysis machines, which make use of this osmosis effect to act as artificial kidneys for extracting substances like urea from blood.

The osmotic pressure difference between salt water and fresh water is equivalent to the hydrostatic pressure of a column of water about 700 feet (213 meters) high!

The potential for generating electric power using gradients in salt concentration is great because of this large osmotic effect. Conceptually, all one needs to do is find a suitable membrane material, build a system for using osmosis to raise diluted ocean water to a height above that of fresh water, and then let this ocean water flow back down through a turbine to generate electric power in the same way that hydroelectric power plants do. But accomplishing this result is extremely difficult in practice. To begin with, suitable membrane materials must be thin to be usable, in order to allow rapid water transport. Thin plastic membrane sheets can't hold back very much hydrostatic pressure, even when supported by porous supports that let the water through while backing up the membrane. Also, membranes exposed to seawater have trouble with fouling and easily become blocked. To get around the need for a membrane, it is possible to imagine making use of the difference in vapor pressure of fresh water versus salt water at the same temperature.

Because of the salt it contains, the vapor pressure of salt water is less than that of fresh water. That's why salt water boils at a slightly higher temperature than does fresh water. At atmospheric pressure the boiling point of ocean water is about 103.7 °C, and at 100 °C its difference in vapor pressure compared with that of fresh water is about two pounds per square inch, which is about eight times the pressure difference that Claude had to work with. However, at lower temperatures, this vapor pressure difference is much smaller. At about 60 °F (15 °C), for example, the difference in vapor pressure between fresh water and ocean water is only 0.14 pounds per square inch, which is even less than the pressure difference that Claude had to work with. But at least this method does not require a membrane. One possibility that does not seem to have been evaluated is using solar water heating to raise the temperature of the fresh water to increase the difference in pressure between it and ocean water. Of course, using warmed fresh water to heat ammonia, chlorofluorocarbons, or hydrochlorofluorocarbons would give even higher pressures to work with, but doing so requires the use of heat exchangers to keep the ammonia or chlorofluorocarbons from escaping. Simply heating fresh water doesn't have this problem, and it increases the difference in vapor pressure that naturally exists between ocean water and fresh water. This concept needs further evaluation to determine whether or not, in the end, it might just be easier to use solar cells to produce electricity directly.

Saltier water makes all salinity gradient power schemes look better. For this reason several conceptual methods of generating electric power have involved the Dead Sea, which is much saltier than the ocean. So much so, in fact, that the hydrostatic head that could be developed between fresh water and Dead Sea water is 20 times greater than that between fresh water and ocean water. The Great Salt Lake in Utah is also much saltier than seawater, but its salinity varies seasonally. The saltiness of the Great Salt Lake can be as much as seven times that of the open ocean. None of the Dead Sea schemes have been put into practice, but very salty water is also produced in the process of water desalination. It is at least conceivable that the very salty wastewater thrown away by a desalinization plant could be used with normal seawater rather than fresh water to produce electric power.

It is also possible to build battery-like electric power sources, called *dialytic* batteries, that use fresh and salt water to generate electricity. Such devices use the

The Dialytic Battery

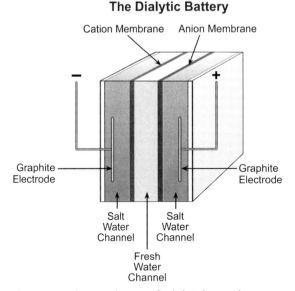

Figure 15.2 Schematic drawing of a dialytic battery that uses the electrochemical potential difference (about one-quarter of a volt) that can be developed between salt water and fresh water by means of semipermeable membranes that allow selective passage of either sodium ions (Na^+) or chlorine ions (Cl^-).

electrical potential that is developed across pairs of membranes exposed to water having different salt concentrations. The resulting device needs membranes, but instead of transporting water, these membranes transport ions. An ionic substance such as salt (NaCl) dissolves in water, and when it does, the sodium and chlorine atoms separate and form electrically charged sodium ions, Na^+, and chlorine ions, Cl^-. In a dialytic battery, one membrane allows only positively charged ions (cations), such as Na^+, to pass through, and the other allows only negatively charged ions (anions), such as Cl^-, to pass through. A schematic drawing of this arrangement is shown in Figure 15.2.

In Figure 15.2, the cation membrane has a saltwater channel on its left side and a freshwater channel on its right side. On the other side of the freshwater channel sits the anion membrane, with a second saltwater channel on its right side. The electrode on the cation membrane side becomes negatively charged, and that side is the battery anode. The electrode on the anion membrane side becomes positively charged, and that side becomes the battery cathode. Other electrochemical reactions occur near both electrodes, just as they do near the positive and negative terminals of automobile batteries. Unlike automobile batteries, however, the dialytic battery is continuously recharged as salt water and fresh water are supplied to it. In this way, a single-cell dialytic battery can produce a small voltage (about one-quarter of a volt), and the net result is that the salinity of the salt water is reduced while that of the fresh water is increased. It is obvious that a large number of dialytic batteries are necessary to achieve reasonably high voltages. A major problem with dialytic batteries is fouling. Ocean water is salty, but it also contains a lot of living things, and anything immersed in seawater tends to become covered in slime. So it is with the membranes in a dialytic battery.

One big advantage of this battery-like approach is that it is not a heat engine as OTEC systems are and is not limited by Carnot efficiency considerations. Such a dialytic battery system was apparently built in the Russian city of Vladivostok several years ago and operated for perhaps three years. Presumably, the membrane used in this system also had fouling problems. Right now, plans are afoot in the Netherlands for tests of a 50,000-watt salinity power-generating station using dialytic batteries.

It appears at present that OTEC power is closer to becoming reality on a large scale than is ocean salinity gradient power. But in the long run, salinity gradient renewable power has a big advantage over ocean thermal energy generation. This advantage involves the fact that rivers running into the ocean represent concentrated renewable sources of power. OTEC power plants, on the other hand, require bringing cold ocean water and warm ocean water together, and doing so requires a lot of plumbing. Location is critical for both ocean thermal and salinity gradient systems, and there are more places where rivers run into the sea than where cold and warm ocean water are close to each other and also close to land.

Many rivers run into the sea, and thus the potential for generating power from salinity gradients is large. Of course, tidal effects can lead to brackish water near the mouths of rivers that empty into the ocean, and in such cases it might be necessary to bring fresh water from further upstream to maximize power production. In the final analysis, of course, it is solar energy that drives renewable hydroelectric power, OTEC power, and freshwater versus saltwater power-generating systems. Salinity gradient power offers the possibility, at least theoretically, of reproducing the equivalent power output of all hydroelectric power systems. This possibility makes research on salinity gradient power systems worthwhile. Presumably the membrane fouling problems of dialytic batteries can be solved eventually. Meanwhile, however, the engineering of OTEC thermal gradient systems, like the ones attempted by Georges Claude, may be approaching practicality for powering communities on remote islands. Unlike solar cells, OTEC plants work day and night.

What OTEC development needs now is another visionary like Georges Claude. In the early 20th century, the technology Claude had available was rudimentary compared to what is available now in the early 21st century, and fossil fuels cost a lot more as well. Deep pockets combined with serious engineering might allow OTEC to become a usable source of renewable electric power. Power from ocean salinity differences is probably going to take longer to become practical, but it does have the great advantage that experiments can be done on a small scale. The scale size needed to carry out pilot tests and experiments is important, as the history of the development of fuel cells demonstrates. Fuel cells are completely different in principle from OTEC power plants and are more like salinity gradient power systems. We will look at fuel cells in the next chapter.

Recommended Reading

D'Arsonval, Jacques-Arsène. "Utilization des forces naturelles. Avenir de l'èlectricite." *Revue Scientifique* (1881): 370–372.

Loeb, S. "Production of energy from concentrated brines by pressure-retarded osmosis," *Journal of Membrane Science* 1 (1976): 49–63.

Pattle, R. E. "Production of Electric Power by Mixing Fresh and Salt Water in the Hydroelectric Pile." *Nature* 174 (1954): 660.

Pittman, Jr., Walter E. "Georges Claude's Magnificent Failure." *Environmental Review*. 6.1 (Spring, 1982): 2–13.

Norman, R. W. "Water Salination: a source of energy." *Science* 186 (1974): 350.

Weinstein, John N. and Frank B. Leitz. "Electric Power from Differences in Salinity: the Dialytic Battery." *Science* 191 (1976): 557–559.

16
Fuel Cells: Hydrogen, Alcohol, and Coal

> ... *I have given an account of an experiment in which a*
> *galvanometer was permanently deflected when connected with*
> *two strips of platina [platinum] covered with tubes containing*
> *hydrogen and oxygen.*
>
> W. R. Grove, *On a Gaseous Voltaic Battery*, 1842

Fuel cells are devices for reacting hydrogen directly with oxygen to produce water and electricity without combustion. This means that they are not heat engines, and thus are not constrained by Carnot's efficiency limit, meaning that they can deliver as electric power a greater fraction of the energy in their fuel than internal combustion engines or any other heat engine can. Fuel cells offer the possibility of another source of power for driving cars, producing only water vapor, not carbon dioxide, as their exhaust. They also can be fueled with alcohol, which acts as a carrier of hydrogen for the cell. High-temperature fuel cells also can be fueled with carbon monoxide and even, at least theoretically, directly with carbon ions. They have a history more than a century long, so let's review how these devices have been developed over time and what they might do to help solve our energy, petroleum, and climate problems.

Fuel Cells and Hydrogen

Fuel cells convert chemical energy into electrical energy without burning anything. In their different forms, they can accomplish this feat with carbon, hydrogen, or many other fuels, such as alcohol or methane. The discovery that electricity could turn water into hydrogen and oxygen was a focus of scientific interest in the 1830s when the British lawyer-turned-physicist William Grove became interested in working this process the other way around, by combining hydrogen and oxygen to produce electricity. His experiments, published in 1839 and 1842, involved the insertion and sealing of platinum wires through the closed ends of test tubes filled with a sulfuric acid solution and inverted into a beaker of this same acidic solution. Using a chemical battery, he then electrolyzed water to produce hydrogen and oxygen in separate glass tubes. By connecting a number of these tubes together,

Energy Demand and Climate Change: Issues and Resolutions. Franklin Hadley Cocks
© 2009 WILEY-VCH Verlag GmbH & Co. KGaA, Weinheim
ISBN: 978-3-527-32446-0

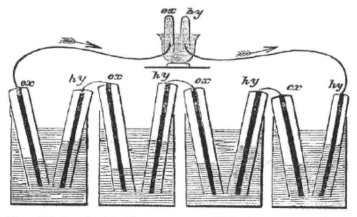

Figure 16.1 Grove's original "gaseous battery" design from his 1842 publication. This figure shows a series of large glass tubes fitted with platinum wires. When electrified by a battery, alternate tubes can be filled with hydrogen and oxygen gas. When these gas- filled tubes are connected to a smaller pair of similar tubes completely filled with electrolyte, they themselves electrolyze water and produce hydrogen and oxygen, thereby proving that they generated electricity.

he was able to use the electricity produced by his "gaseous voltaic batteries" to electrolyze water again. The illustration he used to explain his discovery is reproduced in Figure 16.1.

Grove discovered that once a chemical battery had produced hydrogen in one tube and oxygen in another, a voltage could be measured between the platinum wires in these tubes. By using a sensitive voltmeter called a galvanometer, a voltage could be detected after the chemical battery was disconnected. Although the currents produced by his "gaseous voltaic batteries" were minuscule, the voltage he measured, when no current was drawn from it, was comparable to that of current fuel cells (about one volt per individual cell).

Because the currents generated by Grove's device were dramatically smaller than those generated by chemical batteries, almost no use was made of his discovery until the 1880s. In 1882 the British physicist Lord Rayleigh increased the practicality of the Grove cell by using platinum gauze on the surface of the electrolyte rather than platinum wires, thereby increasing the area of platinum surface, and was able to produce "an inferior, but still considerable current."

Extending Rayleigh's work, Ludwick Mond and Carl Langer in 1889 constructed even more effective devices and coined the name "fuel cell." In their cells the sulfuric acid, rather than being a free liquid, was contained in porous, non-conducting solids such as unglazed earthenware or pasteboard covered with perforated platinum or gold foil coated with platinum powder. The current generated in fuel cells using platinum powder instead of solid platinum was far higher than that obtained by either Grove or Rayleigh because the surface area per gram of fine platinum powder is much higher than that of solid platinum. The fuel cells of the Mond and Langer design are considered the forerunners of modern cells.

The English engineer Francis T. Bacon became interested in fuel cells in the 1930s and by 1955 had developed a system that used an aqueous solution of potassium hydroxide as the electrolyte. Because he used potassium hydroxide rather than sulfuric acid, Bacon was able to replace platinum with nickel for the electrodes, thereby lowering the cost, but nickel catalysts had to be operated at elevated temperatures to be effective. These fuel cells worked well enough to be used on the *Apollo* missions to supply electric power. The water fuel cells make by the reaction of hydrogen with oxygen to produce power was pure enough to be used for drinking by the astronauts.

In part because platinum will catalyze the reduction of sulfuric acid at temperatures above 100 °C, thereby impairing cell performance, fuel cells based on hot solutions of phosphoric acid were developed in the 1970s. These elevated temperature cells produced more electricity than low temperature cells, and a prototype 11-MW phosphoric acid fuel cell power plant was built in 1991 in Japan and operated at efficiencies over 41%.

In 1960 work began on fuel cells that could use a solid polymer as the electrolyte rather than either sulfuric or phosphoric acid. These cells used a polymer that is a superionic conductor of hydrogen ions, that is, a polymer permitting hydrogen ions to move very easily. These solid polymer fuel cells operated at room temperature and supplied the onboard power for the *Gemini* space missions. Such fuel cells are termed proton-exchange membrane fuel cells or polymer electrolyte membrane fuel cells. They offer many advantages for use in powering vehicles, not the least of which is that they can operate at room temperature. They are the fuel cells of choice for powering cars.

Considerable engineering and scientific effort is being expended to use fuel cells as replacements for the internal combustion engine in passenger cars. But fuel cells are expensive compared with gasoline engines. Besides being expensive, another problem with fuel cells is supplying the hydrogen they use: both the hydrogen onboard the vehicle and the network of hydrogen stations needed for fill-ups.

Fuel Cell Efficiency

Because a fuel cell can react hydrogen and oxygen to produce water without combustion, electric power is generated without burning something to boil water and produce steam to generate mechanical work to spin an electric generator. A fuel cell is not a heat engine, so it is not limited by the Carnot efficiency limit discussed in Chapter 9. Its theoretical efficiency for converting the so-called free chemical energy of its fuel into electricity is actually 100%. Sometimes, specialists use a different definition for fuel cell efficiency, such as the ratio of the electrical energy it can generate divided by the heat its fuel could be burned to make. In spite of the high theoretical efficiency fuel cells offer for converting chemical energy into electricity, there will still be electrical resistance losses. Even electric motors are not 100% efficient, and there are also resistance losses within the fuel cell itself.

Nevertheless, vehicles powered by fuel cells are able to achieve overall fuel-to-wheel efficiencies of about 50%, twice as high as the 25% efficiency currently attained by internal combustion–powered vehicles. In the next section we will review the factors that limit efficiency.

Fuel Cells and Cars

The fuel cells of choice for powering cars use polymer electrolyte membranes. But even in fuel cells using a polymer (plastic) membrane instead of acid as the electrolyte, moisture is still required, because it is at the fuel–water–platinum–solid electrolyte interface that electrochemical reactions occur. The hydrogen and oxygen cannot be dry. On the other hand, the cell produces water vapor on the oxygen electrode side, and if this water is not removed, the cell becomes "flooded," which means that the water layer on the electrodes thickens, and the reaction is slowed down by the need to diffuse the oxygen through it.

The electrical resistivity of the membrane is extremely important in cell operation because all the current through the cell must pass through this membrane. The power loss in the membrane is due to its electrical resistance. This resistance causes heating as the current flows through the membrane. This heat must be removed. Cooling is as important for fuel cells as it is for internal combustion engines. The energy loss resulting from this electrical resistance decreases the energy that is available at the fuel cell terminals. Another major source of energy loss is so-called polarization resistance. Polarization resistance occurs when the concentrations of dissolved oxygen and hydrogen become reduced near the surface of the platinum, if power is drawn from the cell faster than diffusion can supply fresh hydrogen and oxygen.

Hydrogen–oxygen fuel cells produce only pure water as their waste product, not carbon dioxide. Even if used on a massive scale, the effect of fuel cell exhaust on total atmospheric humidity would be trivial compared with the water vapor added to the air by the oceans. It's possible that, in large cities awash with fuel cell cars, the added water vapor from fuel cell exhaust might be noticeable, but compared with smog, water vapor is pretty benign.

With these advantages, why haven't fuel cells been used in cars long before now? The answer is that even a century after Grove invented them, fuel cells still couldn't come close to producing the same energy per pound as internal combustion engines could. It took decades of research and development before they were ready to supply electricity to the *Apollo* spacecraft. But fuel cells can now produce power levels equivalent to gasoline engines of about the same size. How do they do this?

Unlike internal combustion engines, which can burn a variety of fuels, such as gasoline or methanol (or hydrogen) or even peanut oil in diesel engines, current vehicle fuel cell use hydrogen. This hydrogen can be supplied in pure gaseous form or extracted from another fuel, such as alcohol or natural gas (CH_4). Producing hydrogen from hydrocarbons is called reforming, but this process also produces CO_2. Reforming of natural gas is the current industrial method for bulk hydrogen production, and a reforming step is inherently part of an alcohol fuel cell. In fact, breathalyzers are nothing more than small fuel cells operating by electrochemically releasing the hydrogen contained in the alcohol in a driver's breath. But in the end, it is hydrogen reacting with oxygen that determines the breathalyzer reading.

Hydrogen reacts at its interface with water and an oxide-free bare metal surface, usually platinum. Platinum is one of the few metals that normally do not have an oxide. To help the reaction go, the water can be acidified so that it contains a lot of hydrogen ions. Grove's cell used sulfuric acid for this purpose. Hydrogen gas dissolves into the acidified water, reacting on the platinum surface to produce hydrogen ions (hydrogen atoms missing their electrons) and free electrons. These electrons can then leave the cell via the negative terminal. The hydrogen ions in turn diffuse through the electrolyte and consume electrons as they react with oxygen to produce water and a positive voltage on the oxygen side of the cell. The hydrogen ions in the original Grove device had to diffuse through a very long electrolyte path to get over to the oxygen and, worse yet, the hydrogen-acidified water–platinum interface was small. That's why the amount of electricity he could produce was so tiny.

As described, modern fuel cells use extremely fine platinum powder particles, called platinum black, not bulk platinum. Platinum black is just platinum powder that has a particle size smaller than the wavelength of light. It can be produced by slowly neutralizing an acidic solution containing dissolved platinum. Surface area per gram increases as particle size becomes smaller, and the platinum powder used in fuel cells is so small that it can give 40 square meters of platinum surface area for each gram of platinum. In this way, the hydrogen-acidified water–platinum interface is thousands of times larger than it was in Grove's fuel cells. Also, the proton-exchange membrane is very thin, about the same thickness as the plastic used in trash bags, and has a low resistance to the movement of hydrogen ions. Solid polymer plastic electrolytes and the great surface area of platinum black are key factors contributing to the success of automotive fuel cells. The schematic design of a modern vehicle fuel cell is illustrated in Figure 16.2.

Cars typically need about 100 horsepower. How much power can a fuel cell produce? A single cell generates only about one volt. Power (P) in watts is the current (I) in amperes times the voltage (V) in volts: $P = IV$. One horsepower is equal to about 746 watts, so a single fuel cell running at one volt would need to produce 746 amps to generate one horsepower. Supplying this many amps requires a cell with an electrode size of about two square feet (0.18 square meters). To power a car with 100 horsepower from a one-volt single fuel cell would require 74,600 amps, which is unreasonably high. More practically, multiple one-volt cells in series (a fuel cell stack) can deliver the needed power at greatly reduced current levels because power equals voltage times current. Packed closely together, like the plates in a car battery, the voltages of each individual cell add together. The total volume of a 100-horsepower fuel cell stack is about the same as a typical gasoline engine of the same horsepower. One potential problem: if one plate

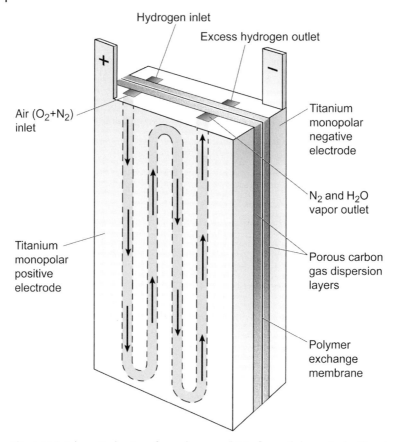

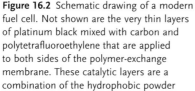

Figure 16.2 Schematic drawing of a modern fuel cell. Not shown are the very thin layers of platinum black mixed with carbon and polytetrafluoroethylene that are applied to both sides of the polymer-exchange membrane. These catalytic layers are a combination of the hydrophobic powder polytetrafluoroethylene mixed with carbon black and platinum black to increase the gas–water–platinum contact area. A large gas–water–platinum contact area increases the current the cell can supply to thousands of times more than that of Groves' original cell shown in Figure 16.1.

in the stack fails, the whole stack fails, just as a chain is only as strong as its weakest link.

A lot of hydrogen is required to keep a fleet of fuel cell cars operating. Building a national infrastructure of hydrogen fueling stations will be a monumental undertaking. And to carry hydrogen onboard, automobiles powered by fuel cells have to store it somehow. So let's look at hydrogen storage next.

Storing Hydrogen

Storing the fuel for a gasoline engine is easy because gasoline is a liquid at room temperature. Storing the hydrogen for a fuel cell is not so easy, because hydrogen is a gas at room temperature. The volume of one pound of hydrogen (0.454 kilograms) at atmospheric pressure is about 1500 U.S. gallons (5678 liters), which is hopelessly large. Hydrogen only becomes liquid at around −423 °F (−257 °C), which makes liquid hydrogen difficult to produce and store. Even as a liquid it's still very light, less than one-tenth the density of gasoline, but pound for pound hydrogen can produce more than three times the energy that gasoline can. And because fuel cells are so efficient, one pound of hydrogen can drive a car six times as far as a pound of gasoline. But even so, it's not easy to store many pounds of the stuff.

A one-gallon volume (3.785 liters) of hydrogen gas pressurized to 10,000 pounds per square inch (69,000 kilopascals) can drive a fuel cell car for only a relatively few miles. It's easier to store larger amounts of pressurized hydrogen at a given pressure as the temperature drops. With a jacket of liquid nitrogen at −321 °F (−196 °C), the same one-gallon volume of hydrogen at 10,000 psi could propel a fuel cell–powered car about the same distance as a gallon of gasoline propels a regular car. Of course, as compressed hydrogen warms up, it must be vented, or the pressure in the tank will increase. This venting can be done via platinum catalyst gauze to react the hydrogen gas flamelessly with air so that only water is released. Pressurized hydrogen ensures that a fuel cell–powered car parked at an airport would still have some hydrogen left when you return and all the liquid hydrogen has turned into hydrogen gas. If steel were used to contain the pressurized hydrogen, the tank would be very heavy. Fortunately, carbon fiber–reinforced fiberglass tanks can withstand high pressures and weigh only about 30% as much as steel tanks. Even so, the volume and mass end up much greater than that for gasoline tanks. Cars powered by fuel cells are going to be short on storage space for luggage.

Besides pressurization or cooling until it becomes a liquid, hydrogen can also be stored as a powdered metal hydride, such as magnesium dihydride (MgH_2). Such hydride compounds are potentially rechargeable with hydrogen. Even though magnesium isn't especially dense, weighing only 1.45 times as much as water, about 13 pounds of MgH_2 is needed to supply one pound of hydrogen. Worse yet, MgH_2 must be heated to about 500 °F to release its hydrogen. This release temperature can be lowered a lot by using $MgNiH_2$, but the total mass goes up, as nickel is heavier than magnesium. Lots of other storage methods are being investigated, including adsorption of hydrogen on the surface of light powders such as silica or soot-like carbon black. Hydrogen-rich chemicals such as alanates (compounds like $NaAlH_4$) are also being evaluated. Even the storage of hydrogen under extremely high pressures in tiny hollow glass beads is being considered. It turns out that glass, when heated, releases hydrogen rapidly via diffusion. To date, none of these novel storage ideas has been shown to be as practical as either pressurizing or liquefying hydrogen.

If batteries were 10 times better at storing electricity than they are, fuel cell technology might not be needed. But after a century of effort, lead–acid battery-powered cars still can't go very far between charges, and recharging a battery takes a lot of time. As was discussed in Chapter 11, lithium and lithium-ion batteries are far lighter and store more energy per kilogram than lead batteries. It appears possible to produce a lithium-ion battery electric car that can go more than 300 miles between charges. Lithium batteries, however, have an unfortunate tendency to burn during heavy charging or discharging. Presumably, this problem can be solved, but right now the potential for lithium batteries to catch fire during charging or discharging is a major problem. Recent work on improved membrane materials for lithium-ion batteries should help make lithium batteries better.

Stored hydrogen might also explode in an accident, but gasoline presents a similar hazard. A car must carry a lot of high-energy fuel to go 300 miles between fill-ups. Because hydrogen vapor is lighter than air, it floats upwards and dissipates if released. Of course, hydrogen also burns, as it did in the *Hindenburg* disaster. Whether gasoline or hydrogen is used, the energy in fuel will always be a hazard.

Producing Hydrogen

Where can we get all the hydrogen needed to run millions of vehicles? Lots of so-called "well-to-wheels" evaluations have been made of hydrocarbon fuels as sources of hydrogen. The current industrial process for reforming natural gas (CH_4) makes little sense for producing the hydrogen to power fuel cell cars on a large scale. It would be simpler just to burn natural gas in an internal combustion engine, as many taxi fleets already do, even if fuel cells are more efficient. A better long-term approach is to make hydrogen from water. The Earth has oceans of it. All we need is the energy to extract this hydrogen.

Electricity, of course, can be used to separate hydrogen and oxygen from water. But producing hydrogen by using the electric power from coal-, gas-, or oil-fired generators doesn't do a thing to reduce global CO_2. In fact, electrolysis of water using electricity from a power plant would be the worst possible way of generating hydrogen on a large scale. The inefficiency of burning something to boil water to make steam to spin a turbine to generate electricity, and the losses in transmitting this electricity to where it's used, would mean adding enormously to the CO_2 burden of the atmosphere, thereby defeating the purpose of using hydrogen and fuel cells in the first place. Fortunately, it turns out that there are several ways of generating hydrogen without producing CO_2.

Technologies for Hydrogen Production

Only CO_2-free methods of producing electricity make long-term sense for the large-scale electrolysis of water. Electricity from solar cells, nuclear reactors, or any other source can be used to separate water into hydrogen and oxygen or to charge the batteries in electric cars. Most of the world's electricity is produced by burning fossil fuels. In 2004, about 19% was generated by hydropower, 16% came from nuclear power, and less than 3% was produced by solar cells, wave, tidal, and wind power combined. To power more than a billion vehicles driving billions of miles per year, enormous quantities of hydrogen would be necessary. It would be nice to do this without producing CO_2.

It takes energy to separate hydrogen from oxygen, but the energy needed decreases as the temperature of the water is increased. In fact, at high temperatures (e.g., above ~4000 °F or 2200 °C), water begins to decompose into hydrogen and oxygen. But 4000 °F is a very high temperature indeed, so chemical engineering processes have been developed that allow hydrogen to be separated from oxygen at much lower temperatures by means of a series of chemical steps.

For example, using a chemical cycle involving calcium, bromine, and iron, thermochemists have lowered to less than 1500 °F (816 °C) the temperature needed to produce hydrogen from water. Lower process temperatures make separation of hydrogen from oxygen with heat a lot more practical. It may be possible to combine high temperatures with chemical engineering to produce both electricity and hydrogen simultaneously using energy supplied by a nuclear reactor. Some have estimated that the resulting hydrogen might cost the equivalent of gasoline at 36 cents per gallon (10 cents a liter)! This cost is especially low because the plant would also generate electricity and sell it, since none would be needed for hydrogen production. Of course, one then has the difficulty of storing the radioactive waste from the nuclear reactor. But at least the problem is shifted away from CO_2, global warming, and gasoline.

Besides technologically advanced schemes, it is possible to produce hydrogen from just water and coal. In times past, burning coal with water vapor and a reduced air supply yielded "producer gas," a mixture containing carbon monoxide and hydrogen. Until the "Big Inch" gas pipeline from Louisiana reached New England in the 1950s, producer gas (not natural gas) was used for heating and cooking in the northeast United States. In a process being investigated at Los Alamos National Laboratory, coal is burnt with water vapor, limited air, and a magnesium-bearing mineral, such as serpentine, to produce magnesium carbonate plus hydrogen, without adding CO_2 to the atmosphere. And vast deposits of serpentine and other magnesium silicate minerals are available worldwide. Since 2001, many conferences have been held to discuss so-called carbon "sequestration" methods for dealing with CO_2 produced by coal-fired power plants. Carbon dioxide can be stored for long times either by forming carbon-containing minerals or by injecting it into oilfields or into the deep ocean. As discussed in Chapter 4, CO_2 dissolves into oil and lowers its viscosity, thereby increasing oilfield production rates.

Certain materials, for example, cerium oxide, act as catalysts when dispersed in water to use some of the 11% of the energy in the ultraviolet portion of the solar spectrum for splitting water directly into hydrogen and oxygen. Unfortunately, such catalysts are not very effective and, to date, are not even close to being practical. Currently known photovoltaic–photoelectrochemical processes that simultaneously produce hydrogen and electricity, or biological methods involving certain algae that produce hydrogen in the course of photosynthesis, are not practical yet either. Photodecomposition of water is discussed in Chapter 19.

Fuel Cells and Coal

Not all work on fuel cells involves hydrogen. The German chemist William Ostwald (winner of the Nobel Prize for chemistry in 1909) proposed in 1894 that electrochemical oxidation of coal had the potential to convert a greater percentage of this chemical energy into mechanical or electrical energy than burning coal to boil water and power a steam engine. He was right, and it still does have that potential. In Ostwald's day, steam engines had an efficiency of only about 10%. Following up on his suggestion, various workers constructed cells consisting of a platinum crucible, molten sodium hydroxide, and lumps of coal submerged in the molten sodium hydroxide and connected via a platinum wire. Such fuel cells were then termed "indirect" fuel cells because, although coal was consumed, the cell itself turned out to operate on hydrogen and oxygen, the hydrogen being produced by the reaction of the coal with the sodium hydroxide. The end results were far from satisfactory, and in the end, liquid electrolytes for fuel cells fed with coal were abandoned in favor of solid electrolytes, especially zirconium dioxide (zirconia) stabilized with calcium oxide (lime).

To avoid a molten electrolyte, solid oxide fuel cells use the fact that calcium oxide–stabilized zirconia, when heated to around 1500 °F (about 800 °C), conducts electricity reasonably well because of the movement of oxygen ions. Near the end of the 19th century, the ceramic material zirconia was used to make incandescent lights for theatrical performances. These lights were more stable than arc lamps and much brighter than the weak, carbon-filament lights Thomas Edison had invented. High electrical currents heated stubby calciumoxide-zirconia cylinders to extremely high temperatures, and they produced a light that was extremely bright. The phrase "in the limelight" originated from these stage lights and theater lighting that used pure calcium oxide rods heated by the flame from a burning oxygen–hydrogen torch, because of the calcium oxide (lime) involved. In any event, the ability of zirconia to conduct electricity by moving oxygen ions instead of electrons makes it possible to oxidize carbon monoxide to carbon dioxide and produce electricity in a fuel cell whose basic fuel is coal. All these zirconia fuel cells used platinum to collect the electric currents they produced by oxidizing carbon monoxide to carbon dioxide electrochemically instead of by combustion.

An enormous effort has been made over many years to use high-temperature fuel cells based on zirconia (not plastic!) as an electrolyte to increase the efficiency of coal-fired power plants. It turns out that zirconia, unlike most ceramics, transports oxygen ions very well at high temperature. In most such high-temperature fuel cell schemes, coal is first burned with insufficient air to produce carbon monoxide. This carbon monoxide is then oxidized to carbon dioxide via the transport of oxygen ions across the superionic zirconia electrolyte to produce electricity in the usual fuel cell manner, after which the resulting hot carbon dioxide is used to boil water and generate more electricity, thereby improving the overall efficiency of the power plant. Recently, it has been suggested that a solid electrolyte for carbon ions would enable the direct oxidation of carbon to CO_2 without any combustion at all, but so far such a carbon-ion electrolyte remains only a dream.

Presently known carbon-ion solid electrolytes don't transport carbon ions very well.

The combination of a high-temperature carbon fuel cell with a normal power station to produce a hybrid fuel cell–boiling water power plant is not yet commercially viable, even after decades of effort. Instead, what has become commercial is the use of hydrogen fuel cells based on the old W. R. Grove concept. Even these cells took a long time to become practical.

Fuel Cells and Alcohol

In addition to hydrogen and carbon or carbon monoxide, fuel cells can also operate using hydrocarbon fuels such as alcohols, especially methanol or ethanol, which act as carriers for hydrogen. In such cells, the electrochemistry gets very complicated. To avoid these electrochemical complexities, it is possible to use reforming processes to thermochemically strip hydrogen away from methanol (CH_3OH) or ethanol (C_2H_5OH) and then use this hydrogen in a traditional hydrogen fuel cell. However, the reforming process has its own difficulties, one of which is that fairly high temperatures are required. Instead of reforming methanol or ethanol, it is possible to use these liquid fuels directly in a fuel cell; such cells are quite logically called direct methanol or direct ethanol fuel cells. The difficulty is that the complex electrochemistry involved in stripping the hydrogen from the methanol or ethanol means that generating a high level of power requires a large alcohol fuel cell size, much bigger than for hydrogen fuel cells. Direct ethanol fuel cells are much worse than direct methanol fuel cells, and neither works as well as hydrogen fuel cells do. Direct methanol cell cars might possibly be feasible, but direct ethanol fuel cells seem to be virtually impossible, at least currently. Using methanol in a direct fuel cell would solve the problem of storing hydrogen. Also, gasoline stations exist all over the world and could be easily modified to include methanol in addition to gasoline. Finally, methanol can be made from wood, and so-called "wood alcohol" has long been available. If the methanol is made from trees, most of the CO_2 released by direct methanol fuel cells would have come originally from the air. In a rational world, it would seem appropriate to establish a serious research effort on the trees-to-methanol concept and direct methanol fuel cells for vehicles, but not very much work is currently being done on this possibility.

What Happens Now?

The changeover from gasoline to hydrogen or biofuels (see Chapter 12) for powering cars will require a huge effort, nationally and internationally, major changes in numerous laws, and many hundreds of billions of dollars in cost. If CO_2 and global warming were the only driving forces, it might take a very long time before such an effort was made. But if gasoline starts costing $10 a gallon and only $50 worth can be bought at a time, this changeover will be greatly speeded up. At high enough prices gasoline would also probably be produced in quantity from coal, even though this coal-to-gasoline process produces a lot of carbon dioxide. One way or another, the way in which fuel for tomorrow's cars is obtained and used is going to change.

We will face enormous transportation problems in the years ahead, and all solutions will have their special difficulties. There is no such thing as a free lunch. In the end, the final choice of what to do almost certainly will be decided on the basis of cost, availability, and environmental concerns, however these are measured. Fuel cell–powered cars could ease the painful difficulties posed by the approaching end of the age of cheap oil, but this changeover won't be easy. But it is not impossible either.

Recommended Reading

Brown, L. C., *et al.* "High Efficiency Generation of Hydrogen Fuels Using Nuclear Power." General Atomics Report GA-A24285, December, 2003, prepared under the Nuclear Energy Research Initiative (NERI) program. (See p. 5–1 for estimated hydrogen cost.) See also: http://www.eere.energy.gov/ hydrogenandfuelcells/pdfs/32405d.pdf

Hoogers, G. ed. *Fuel Cell Technology Handbook.* Boca Raton, FL: CRC Press, 2003.

Larminie, James and Andrew Dicks. *Fuel Cell Systems Explained*, 2nd ed. New York: John Wiley, 2003.

Vielstich, Wolf, A. Lamm, and H. Gasteiger, eds. *Handbook of Fuel Cells: Fundamentals Technology and Applications.* 4 v. New York: John Wiley, 2003.

17
Magnetohydrodynamics and Power Plants

Magnetohydrodynamics: The physics of the motion of electrically conducting fluids in the presence of magnetic or electric fields.

Magnetohydrodynamics (MHD) offers the potential of substantially increasing the efficiency of fossil fuels for producing electricity, thereby reducing carbon dioxide emission.

The interaction between moving, electrically conducting fluids and magnetic or electric fields can be complicated. For example, MHD forces operate in the molten interior of the Earth and generate the magnetic field surrounding our planet. MHD is also involved in the cauldron of the sun and in the magnetic confinement of plasma used to produce nuclear fusion as well as in auroras and solar flares.

MHD power plants can be more efficient than normal plants because they use higher temperatures for power production. In MHD power generation, the fuel is burned, usually using preheated air for combustion, and the resulting high combustion temperatures produce a plasma exhaust that is sent through a magnetic field. In physics, a plasma is an ionized gas that contains a mixture of free electrons and charged ions. In normal power plants, steam turbines spin copper coils in a magnetic field to generate electric current. In MHD power plants, it is a high-temperature, electrically conducting plasma that acts as a conductor moving through a magnetic field, not copper coils. One major difficulty in making MHD power plants practical arises from the need to extract the electric current that the magnetic field induces in the plasma.

Typical power plants operate normally with steam temperatures less than 1000 °F (538 °C) and have efficiencies usually less than 40%. MHD power plants operate at temperatures much higher than this, and could reach efficiencies of 60% or more. Getting the electric power from the plasma into normal wire electrical conductors is one major problem confronting MHD power plant designs. Another is their need for very strong magnetic fields that extend through a large volume. Let's follow the path from fuel to power in an MHD system and see how all this works.

The fuel for MHD power plants can be any combustible material. This fuel is burned in a combustion chamber, where preheated air is used to increase the

Energy Demand and Climate Change: Issues and Resolutions. Franklin Hadley Cocks
© 2009 WILEY-VCH Verlag GmbH & Co. KGaA, Weinheim
ISBN: 978-3-527-32446-0

flame temperature, and the resulting hot plasma is expanded at high velocity through a nozzle. The plasma exiting the nozzle is shot through a very strong magnetic field, about 100,000 times the strength of the Earth's field, to separate its positive and negative charges. To produce the necessary high-strength magnetic fields, coils of a superconducting material must be used. At present, the need for superconductivity typically requires wires made from an alloy of niobium and titanium. Such wires can carry large electric currents without any electric loss, but they have to be cooled to within a few degrees of absolute zero to do so. The niobium–titanium superconducting wires are cooled by super-cold liquid helium that produces temperatures near 4.2 °K (remember, absolute zero, the lowest possible temperature, is 0 °K). The juxtaposition of high-temperature plasma and liquid helium-cold superconductors creates problems. These problems are solved by using extremely effective cryostats to contain the helium and the niobium-titanium superconducting magnets. Superconducting magnets operate just like any other electromagnets, except that, being superconducting, they don't need any energy to maintain the magnetic field once it has been produced.

The high magnetic field strengths needed are similar in magnitude to those used in magnetic resonance imaging (MRI) machines in hospitals to image tumors and other diseases, but the volume over which the field extends is much greater. MRI machines also use superconducting wire made from niobium–titanium alloy wires cooled with liquid helium. It turns out that strong, large-volume magnetic fields are easier to produce with niobium–titanium than with ceramic superconductors, even though ceramic superconductors would only need to be cooled with liquid nitrogen, which isn't as cold as liquid helium and is a lot easier to produce. The level of current that can be handled by niobium–titanium wires is much higher than that for ceramic high-temperature superconductors, and the higher the current, the stronger the magnetic field.

What happens when the high-temperature plasma enters the strong magnetic field produced by the superconducting coils? That's where complexity really begins, and the magic of magnetohydrodynamics comes into play.

Faraday Induction and the Hall Effect

Chapter 8 has already explained Michael Faraday's discovery that moving an electrical conductor, such as a wire coil, in a magnetic field produces an electric current. The plasma moving through the magnetic field also produces an electric current, provided this current has somewhere to go. To complete the electric circuit, the plasma must contact some type of electrode to deliver the electric current. After the electrodes pick up the electric current, a lot of electrical manipulation is needed to convert this direct current into the alternating current that people use in their homes. The real problem involves the electrodes that the supersonic plasma hits as it delivers millions of watts of power. As you can imagine, these electrodes don't last very long. After all, the plasma is also extremely hot and is moving very fast. To achieve a high power density from the

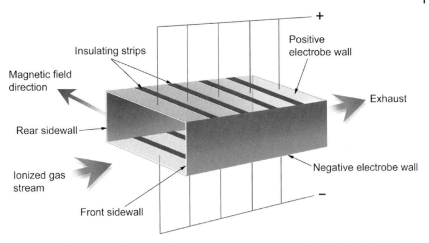

Figure 17.1 Illustration of an MHD electric power generator showing high-speed plasma directed through a channel that has a magnetic field perpendicular to the direction of plasma flow. This configuration produces electric current collected by electrodes on the top and bottom of the plasma channel. MHD generators are capable of producing high levels of power, and the exhaust exiting the plasma channel is very hot and can be fed into the boiler section of a normal power station, thereby substantially increasing the overall efficiency with which electricity is generated per unit of fuel consumed.

plasma, its conductivity must be very high, that is, it must be extremely hot and/or it must be seeded with an element, such as potassium, that increases its electrical conductivity. We're talking rocket exhaust conditions here. You can just imagine what this hot, powerful blast does to the electrodes, especially since the density of power into the electrodes is very high, around 680 kilowatts per square foot (24 megawatts per square meter).

Figure 17.1 is a schematic drawing showing the operation of an MHD power-generating unit. In this unit, the magnetic field is applied across the channel and the plasma is sent at high speed through the channel. The electrical power is collected by electrodes spaced at the top and bottom of the plasma channel. These electrodes are exposed to extreme conditions and they wear out rapidly.

The electrode problem is only partially solved by using high-purity, water-cooled copper to pick up current from the plasma. Another difficulty is the accumulation of slag from the combustion process on the electrodes, increasing their electrical resistance. The problem of the electrodes is doubly critical because these electrodes also must pick up the electric current produced by the Hall effect. In 1879 the American physicist Edwin Herbert Hall found that a current moving in a magnetic field produces a voltage across a direction perpendicular to that of the original current and also perpendicular to the magnetic field. The Hall effect is not the same as Faraday induction. Melding together the electric power produced by Faraday induction and the Hall effect requires a lot of electrical engineering. The interactions between moving, electrically conducting fluids and magnetic fields are very complex, to say the least.

Benefits of MHD Power Generation

The exhaust gas from the MHD plasma channel is still very hot and can be fed directly into the combustion chamber of a normal power plant. The heat from this plasma can be used to boil water and generate additional power using steam in the normal power plant way. The power produced by the MHD plasma adds to the power produced by the steam. A typical power plant currently has an efficiency of 35–37%. With the added power that can be obtained from MHD power generation, total plant efficiency could exceed 60%. In addition, if coal is used as the fuel, sulfur in the coal combines with the potassium that is added to increase the conductivity of the plasma. This sulfur is removed as the stable compound potassium sulfide, which reduces the sulfur content of the exhaust gas that enters the atmosphere. These two benefits, high efficiency and cleaned-up exhaust gas, are two drivers for the development of fossil-fueled MHD power plants. An important obstacle to MHD power plants is the problem of the short lifetime of the electrodes that are needed to collect the current produced by Faraday induction and the Hall effect. Of course, a material that is resistant to damage from the plasma would help in solving the electrode problem. High-temperature materials have been developed for both rockets and jet engines, but these materials have very high electrical resistances compared with copper. The design of low-resistivity copper electrode systems that allow new electrode material to be advanced into the plasma channel as the electrodes erode away, without shutting the plant down, might help.

MHD power plants up to 25 megawatts in capacity have been constructed and operated and have successfully delivered this power into a national power grid. The principle of the MHD method has therefore been proven, but a lot more engineering and materials science work are needed to improve plant lifetime, reliability, and cost. The need for liquid helium to cool the large superconducting magnet is a problem, but closed-cycle helium systems can make the rate of helium loss very small. In addition, certain ceramic materials that become superconducting when cooled by liquid nitrogen ($77\,^\circ$K) also have been developed in the last 20 years, and liquid nitrogen is a lot cheaper than liquid helium. With a lot of research, these high-temperature superconducting materials and liquid nitrogen might be able to replace niobium–titanium and helium. The potential efficiency and pollution benefits MHD offers are important reasons for continuing to develop MHD power generation. Even though fossil-fueled MHD plants would still produce carbon dioxide, they would produce less of it than a normal power plant of the same power-generating capacity because of their higher efficiency.

MHD plants are not a panacea to the world's energy and global warming problem, but they could be a big help if they can be developed to the point that they are able to work reliably and steadily for long times, just as normal power plants do. In recent years, MHD power plant research has stalled, but it should be restarted even though MHD technology has not yet proven to be practical. It has great potential; it is not pie-in-the-sky and should not be neglected in the years to come as it is at present. If all the world's fossil fuel power plants had their efficiencies increased to 60% by adding MHD power generators to them, this step alone would go a long way toward stabilizing the carbon dioxide level of our atmosphere. Surely that possibility deserves serious further investigation.

Recommended Reading

Rosa, R. J. *Magnetohydrodynamic Energy Conversion*. New York: Hemisphere, 1987.

18
Thermionics and the Single-fuel Home

This invention relates to processes and apparatus for converting thermal energy into electrical energy.

George N. Hatsopoulos and Joseph Kaye, U. S. Patent 2,915,652, 1 December 1959

Thermionics is a method for turning heat into electricity without using generators or turbines. It is not a new or renewable source of energy but rather a better way of using a traditional fossil fuel. This technology offers the possibility of utilizing a single primary energy source such as natural gas to provide all the energy needs of homes. And such a thermionic system can be more than twice as efficient as the current means for supplying the energy needed to heat, cool, and light homes, thereby dramatically lowering their carbon footprints. How can thermionics accomplish such a feat? Just what is thermionics, anyway?

In some respect thermionics is akin to thermoelectrics (see Chapter 6), because both technologies can convert heat into electricity, but thermionics is more efficient. Thermionic devices take advantage of the fact that hot surfaces emit more electrons than cold ones do, a principle applied in old-fashioned vacuum tubes. Before the onset of transistors, radios had to "warm up" before they could operate. This warming up time was necessary to heat some of the filaments in the vacuum tubes that made radios work in the 1920–1960 era. Some metals function better than others as emitters of electrons, and as luck would have it, tungsten works very well, especially if it is alloyed with a little thorium. But thorium is somewhat radioactive, which is a bit awkward. The metal rhenium actually works the best, but it is expensive. In any event, besides making radio tubes work, the thermionic effect makes it possible to convert heat into electric power without the need for steam turbines. For power-generating purposes, the rapid reactivity of tungsten with oxygen is a problem, and for this reason the metal rhenium is better, even though it is far more expensive than tungsten.

Thermionic electric generators are heat engines and therefore are hobbled by the Carnot efficiency limit (see Chapter 9), just as steam power plants are. However, the temperatures used by thermionic heat energy converters are much higher than the steam temperatures in power plants. Therefore, the efficiency for converting heat into work using thermionic devices can theoretically be higher than the

Energy Demand and Climate Change: Issues and Resolutions. Franklin Hadley Cocks
© 2009 WILEY-VCH Verlag GmbH & Co. KGaA, Weinheim
ISBN: 978-3-527-32446-0

35% conversion efficiency of most fossil fuel or nuclear power stations. In power plants, the steam and condensation temperatures are around 1000 °F (556 °C) and 100 °F (56 °C), respectively. In thermionic devices the high and low temperatures are around 2420 °F and 980 °F (1600 °K and 800 °K), respectively, giving a Carnot efficiency limit of 1–800 / 1600 or 50% (see Chapter 9). Of course, current thermionic devices don't achieve this great an efficiency in practice, but even so, the important fact remains that the residual heat they reject is at a temperature high enough to power thermal refrigeration as well as home heating. The net result would be a dramatic reduction in the total fuel needed to operate a house year-round compared to the way homes are heated, cooled, and electrified now, without any reduction in living standard. Let's start by describing how thermionic devices operate.

How a Thermionic Converter Works

In the commotion of electrons and atoms inside a heated substance, especially a heated piece of metal, it is not surprising that some of these electrons escape from time to time. The loss of electrons from a heated piece of metal is the thermionic effect. In this way, an electric potential is developed between a heated wire and an unheated one. This thermionic effect was of great interest in the late 19th and early 20th centuries in the days when radio was born. The very earliest radios used crystals of galena (lead sulfide) and pointed wires (called cat's whiskers) as the radio receiver, with bedsprings as antennas, a wire coil as a tuner, and headphones instead of speakers to make the received radio signals audible. These so-called crystal radios did not work very well, but it was not long before the thermionic effect was utilized in vacuum tubes to amplify radio signals enough to power loudspeakers.

Many inventors worked to develop radio tubes, but perhaps none did more to apply the thermionic principal to radio tubes than the American inventor and son of a Congregational minister Lee De Forest (1873–1961). His major invention, the three-filament "Audion," patented in 1907, was the first vacuum tube device that could amplify a radio signal. What followed is history. (Unfortunately for De Forest, he became entangled with unscrupulous financiers and never received the financial reward that his Audion radio tube invention merited). Nowadays, of course, solid-state circuits have replaced vacuum tubes in radios and television. In their heyday, however, production of vacuum tubes of all kinds became the most automated process the world had ever seen.

Of course, thermionic power-generating devices do not involve radio tubes, but just as in radio tubes, the hot surfaces of thermionic devices emit electrons that are collected by cold surfaces. Let's first clear up one point. As a metal surface is slowly heated, a mixture of negatively charged electrons and positively charged ions from adsorbed gases escapes. Metal surfaces are almost always coated with adsorbed gases, especially adsorbed water vapor. As the temperature increases, the metal surface cleans itself up, and the current that is emitted from its surface becomes dominated by electrons. If the heated and the unheated metal surfaces are electrically connected, the resulting electric current between the hot and cold metal surfaces can be used for powering solid-state radios and televisions, as well as many other things. Let's consider next the technology needed to make thermionics practical for generating electric power.

Engineering Thermionic Systems

One fundamental fact: the number of electrons emitted per second from the hot surface of a thermionic converter increases as the square of the temperature times a factor that increases exponentially with temperature. This situation means that the hot emitter surface should be heated to as high a temperature as possible to maximize power generation. If the gas being burned is natural gas (CH_4), then the maximum possible temperature that can be reached is set by the energy released when each molecule of CH_4 is burned with air to produce one molecule of carbon dioxide and two molecules of water vapor. For CH_4, this maximum flame temperature in air is about 3600 °F (1980 °C). Temperatures this hot can make a thermionic electric generator operate effectively, and as a result natural gas is the fuel of choice.

So far, we've only talked about the basics of how thermionic devices work. The application of these basic principles turns up additional important factors, the first of which is the need to coat the heated metal with an easily ionized substance to increase the number of electrons that the heated metal produces via the thermionic effect. Of all the elements in the periodic table, the best ones for losing their electrons easily are heavy alkali metals such as potassium, rubidium, and cesium. In the MHD generators described in Chapter 17, potassium is the element of choice for producing ions because it is by far the cheapest of these three elements, and they use a lot of it. Cesium is the element of choice for thermionic generators because they need only tiny amounts of it to greatly improve performance, and cesium does a better job than either rubidium or potassium. The element francium would work even better than cesium, but every single one of its known isotopes is so radioactive that they eliminate themselves in short order and turn into other elements that don't do thermionic generators any good. Cesium also works better than rubidium or potassium for MHD generators, but MHD power generation would use up such large quantities that the cesium or rubidium cost would be far too high to be practical. In both thermionic devices and MHD generators, the needed alkali metals are so hot that they exist only in the vapor state. But it's fine if the hot metal surface is exposed to cesium vapor, because then the vapor has rapid and repeated contact with this surface.

Repeated contact with cesium increases the rate at which electrons escape from the heated metal plate even though the cesium does not form a solid coating on it. Alloying the metal emitter plate with thorium or dispersing thorium oxide into it also helps, but thorium is radioactive. The cesium vapor's main effect, however, is that in ignited mode the glowing cesium helps neutralize the negative charge resulting from the buildup of electrons in the space between the hot (emitter) plate and the cold (collector) plate. If the cesium vapor is exposed to air, it immediately oxidizes to cesium oxide, and the thermionic device would not work well. For this reason, the gap between the hot and the cold metal surfaces must be evacuated and sealed airtight before the cesium is introduced. During operation, the cesium vapor carries so much current that it actually glows like a neon sign; but of course, this glow cannot be seen because the cesium vapor is sealed between two metal plates. This condition is called the ignited mode, and under these conditions, present-day thermionic cells can produce something like five amperes of current per square centimeter of area, which is hundreds of times more than solar cells produce per square centimeter. Solar cells generating a lot of power require an enormous area of cells. Generating a lot of power using thermionic generators doesn't.

The arrangement needed for a practical thermionic generator is illustrated in Figure 18.1. Two metal plates are needed, one heated and the other cooled. To contain the cesium vapor, which is supplied from an attached cesium reservoir, the gap separating these two plates is fastened around its outer edge with an insulating, vacuum-tight ceramic seal. Importantly, the gap between the two plates must be extremely narrow, around 0.01 millimeters (0.4 thousands of an inch) or

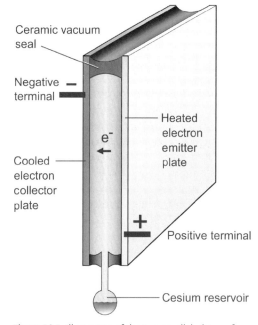

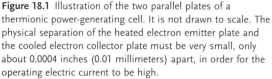

Figure 18.1 Illustration of the two parallel plates of a thermionic power-generating cell. It is not drawn to scale. The physical separation of the heated electron emitter plate and the cooled electron collector plate must be very small, only about 0.0004 inches (0.01 millimeters) apart, in order for the operating electric current to be high.

even less. The high operating temperature of the metal plates and their close spacing cause engineering problems due to warping. This problem of close spacing and warping, which causes the thermionic unit to short out, must be overcome before thermionic converters can become practical. Using ceramic spacers between the plates to keep them separated or developing ceramic electrical conductors that can stand high temperatures and don't warp might solve the narrow-spacing difficulty. All the problems associated with thermionic power generation have not yet been solved. That's why thermionic electrical generators and the single-fuel home are discussed in Part III of this book rather than Part II.

With a small gap between the heated emitter plate and the cooler collector plate, the red-hot metal emitter plate loses a lot of energy by radiation to its partner, the cooler electron collector plate. Heat lost in this way is the major reason why thermionic converters can't be extremely efficient. This heat-transfer problem appears to be resolvable only by increasing the ability of the unheated metal surface to reflect heat radiation rather than absorb it. Up to now this approach has not been especially successful. Obviously, there is a lot of development work still to be done before the single-fuel home becomes practical. It is worth noting, however, that the former Soviet Union produced a number of thermionic power units, which they called the TOPAZ series, for powering satellites. These thermionic power units produced electricity from the heat given off by the radioactive decay of plutonium-238. After the collapse of the Soviet Union, the United States purchased some of these TOPAZ units to study their engineering with a view to using such power sources in U.S. satellites, but in the end, the decision has so far been to stick with nuclear-powered thermoelectric units powered by plutonium-238 as power sources for deep-space missions.

Some might suggest that thermionic converter technology does not need to be developed at all. After all, simply using natural gas to run a motor-generator can produce electricity. However, the exhaust heat from the motor-generator unit is not hot enough to operate a cooling system effectively, while the heat exhausted from a thermionic electric generator is. Generating electricity using thermionics and then using high-temperature waste heat to run an air conditioner or for heating makes very efficient use of the energy in natural gas. That's why thermionics can dramatically reduce the carbon footprint of homes that are electrified, heated in the winter, and cooled in the summer.

In several densely populated cities in the Northern Hemisphere, especially Moscow, the capital city of Russia, waste heat from power plants is used for domestic heating, thereby substantially increasing the efficiency of fuel use. For most power plants in the world, however, the waste heat they produce is just vented to the atmosphere using cooling towers or is discarded into lakes, rivers, or ocean water. For most power plants it is impractical to pipe this low-temperature waste heat over long distances for use in homes or businesses. Using thermionics, it is possible to imagine offering reliable, gas-fired, thermionic power units for domestic and commercial heating, cooling, and electric power applications. In the 1960s the Greek-American engineer and entrepreneur George Hatsopoulos first proposed the concept of the single-fuel home powered by thermionic generators burning natural gas. In addition to being valuable in making maximum use of the energy that natural gas can supply, thermionic units can be scaled to single-home sizes and the choice of whether or not to use them left to economics.

Wide use of such thermionic systems could offer another approach to the waste heat dilemma that is an inevitable consequence of Carnot's efficiency limit for heat engines of all kinds. Building new power plants closer to densely populated areas so that district heating can make use of the waste heat such plants produce is another approach. For suburban homes in northern climates in the wintertime, excess electric power from constantly operating domestic thermionic systems could be fed back into the electric power grid in the same way that excess electric power from solar cells is now. All the engineering needed to feed power back into local electric grids has already been developed and is practical and effective. But thermionic units of modest cost have not yet been developed. The science of thermionics is already known. But it will take a major engineering effort to develop reliable and cost-effective thermionic systems for use in homes.

For one thing, a thermionic cell produces an output voltage of only about half a volt, just about the same as that produced by each solar cell in a solar panel. For this reason, solar cells are usually wired together to increase the voltage produced by the panel. Voltages much higher than 0.5 volts are needed because power is the product of voltage and current. To transmit a lot of power at 0.5 volts means that the current must be very high, and to carry this current, the wires would have to be large and heavy. By wiring solar cells together so that their voltages add up, the current level needed for any given power is reduced and so is the size of the copper wire needed to transmit this power. The same goes for thermionic units, but with one important difference. The current produced by each solar cell is typically less than one ampere. The current produced by each thermionic cell can easily be several hundred times as great. It takes thousands of solar cells to produce the kilowatts of power needed for lighting and cooking in a typical home, but it would take only dozens of thermionic cells.

By using individual home thermionic units to generate electricity, the heat that is not converted into electricity can be used for cooling via the heat-operated refrigerator principle invented in the 1920s by Carl Munters and Baltzar von Platen (see Chapter 6). As energy supplies slowly tighten, the CO_2 level of the atmosphere increases, and the climate changes, the single-fuel-home concept should be seriously considered as one of many approaches toward slowing down global warming and making our supplies of fossil fuels last longer. Just as individual homes can be powered by solar cell systems, so too could individual houses be powered by thermionic systems.

Undoubtedly, the maintenance and operation of thermionic units supplying the heating, cooling, and electric power needs of single-fuel homes will be more demanding than the heating, cooling, and electric power techniques now used. Thermionic technology is not yet ready to go to market, and it will only receive serious consideration as fuel shortages worsen and solutions to the Earth's energy and climate problems are sought with ever-increasing urgency.

Recommended Reading

Hatsopoulos, G. N. and E. P. Gyftopoulos.
   Thermionic Energy Conversion. 2 v.
   Cambridge, MA: MIT Press, 1973, 1979.

19
Artificial Photosynthesis and Water Splitting

> *I am far from thinking that I have discovered the whole of this*
> *salutary operation of the vegetable kingdom; but I cannot but*
> *flatter myself, that I have at least proceeded a step farther than*
> *others, and opened a new path for penetrating deeper into this*
> *mysterious labyrinth.*
>
> John Ingen-Housz, *Experiments upon Vegetables, Discovering*
> *Their Great Power of Purifying the Common Air in Sunshine,*
> 1799

Photosynthesis is the process by which plants convert carbon dioxide, water, and sunlight into carbohydrates and oxygen. If the chemistry that nature uses in plants can be carried out *in vitro* in glass reactor vessels, as opposed to *in vivo* in living plants, the possibility then opens up for using all of the products of this photochemistry for food or fuel production, without diverting any for growth and reproduction. Plants use the energy in sunlight, in the blue, violet, and ultraviolet portions of the solar spectrum, to split atomic bonds in water. This splitting process releases electrons from hydrogen–oxygen bonds and forms energetic hydrogen in chemical complexes that then react through a series of chemical interactions with carbon dioxide molecules to form carbohydrates and oxygen. How plant chemistry works was utterly unknown throughout almost all of human history. The unraveling of the chemical processes occurring in plants only began with the Dutch physician Jon (John) Ingen-Housz (1730–1799).

Ingen-Housz had achieved fame by using live human smallpox virus (not the more benign cowpox virus used later on) to immunize the entire family of Maria Theresa, Empress of Austria. The wealth that accompanied his success with small-pox immunization enabled him to pursue purely scientific studies, including the relationship of animal respiration and plants. Ingen-Housz tried to answer the question of whether plants can use the air breathed out by animals and vice versa. By his own reckoning, he carried out more than 500 experiments involving a number of different plants, including sage and peppermint. He is considered the father of photosynthesis for his discovery that mice survived and did not suffocate when sealed inside jars containing living green plants, but only if sunlight was

Energy Demand and Climate Change: Issues and Resolutions. Franklin Hadley Cocks
© 2009 WILEY-VCH Verlag GmbH & Co. KGaA, Weinheim
ISBN: 978-3-527-32446-0

available. Others had made observations concerning the survival of mice sealed inside jars with plants, but Ingen-Housz was the first to show that sunlight, in addition to the green plants themselves, was an absolute requirement for mouse survival. In modern terms, Ingen-Housz proved that plants in sunlight produce oxygen from the carbon dioxide exhaled by animals.

Molecules of all kinds can be split apart if the temperature is high enough, but the temperature needed to break apart water molecules by heat alone is over 3632 °F (2000 °C). By using a series of smaller chemical engineering steps, the temperature necessary to decompose water into hydrogen and oxygen, without sunlight, has been lowered to around 1472 °F (800 °C). Plants, of course, carry out this process at normal ambient temperatures, but they need the help of sunlight and chlorophyll to do so. In the deep oceans, near volcanic vents where sunlight never shines, plants have developed the ability to break apart volcanic hydrogen sulfide (H_2S) to obtain the energy they need to grow and reproduce in total darkness, but not the ability to use carbon dioxide to produce oxygen. The process of plant growth using H_2S involves an entirely different chemistry than that of plants in sunlight. Plants growing in the dark ocean depths are not green and do not contain chlorophyll. Neither do mushrooms, which also flourish without light, but in their case by consuming the remains of other plants or animal waste. The overwhelming bulk of all plants on our planet rely on sunshine and chlorophyll, or closely related molecules, to live, grow, and reproduce.

Artificial photosynthesis tries to duplicate what green plants in sunlight do, but without involving living organisms. In effect, artificial photosynthesis represents an attempt to develop a series of chemical steps that, combined with sunlight, lower the temperature for the decomposition of water far below that which chemical engineering currently takes to accomplish this feat without sunlight.

One difficulty is the enormous complexity of the chemistry involved in photosynthesis. A lot of progress has been made, but we still don't understand all the details of the "mysterious labyrinth" of photochemistry. Fantastic chemistry is involved in the functioning of both the plant and animal kingdoms. Through an eons-long process, nature has succeeded in developing self-assembling biochemistry of unfathomed complexity. Interestingly, plants preceded animals on the Earth by more than a billion years. Animals need oxygen to live, and the atmosphere of the early Earth was extremely lean in O_2 but rich in the CO_2 that green plants need. Importantly, the high level of CO_2 in the atmosphere of the early Earth gave rise to a major global warming that compensated for the low level of solar intensity during the young life of the sun. By turning CO_2 and H_2O into oxygen and carbohydrates, plants enriched Earth's atmosphere in oxygen and thereby made animal life on Earth possible, just as they make it possible for mice to live in sealed jars together with plants in sunlight.

In nature, the chemical processes in plants are inevitably constrained by the need to keep them alive and able to reproduce. Of the energy in sunlight, 11% has a short enough wavelength to break the bonds in water, a fundamental step in photosynthesis. But in their photosynthetic work, plants use only about 3% of the energy in sunlight, not 11%. It may be that the processes needed to keep plants living, growing, and reproducing constrain the fraction of sunlight that they can use. In artificial photosynthesis, no such constraints would exist, and it is at least conceivable that a greater portion of the energy in sunlight could be utilized for the production of carbohydrates, whether for food or fuel. Artificial photosynthesis might be almost four times better than plants in using sunlight to remove carbon dioxide from the air and turn it into carbohydrates, especially combustible fuels, one of the dreams of artificial photosynthesis and water splitting. However, how to accomplish such a feat is the question. Let's look first at how nature does it.

Plant Chemistry

Most basically, plants combine six molecules of water and six molecules of carbon dioxide and turn them into one molecule of sugar and six molecules of oxygen. How do they do this? The answer involves that wonderful molecule chlorophyll, the green pigment, or its molecular variants, in the leaves of plants. It is noteworthy that magnesium is as important to chlorophyll in leaves as iron is to hemoglobin in blood. Because magnesium is a necessary part of the human diet, telling children to finish eating their green vegetables is good advice.

How does the photosynthesis process work? Briefly, a chlorophyll molecule absorbs a photon of energetic light and loses an electron in the process. This light-activated chlorophyll starts a cascade of chemical reactions that ends by breaking apart a water molecule and releasing hydrogen, which then rapidly combines with other chemical species to form enzyme complexes. Algae and some bacteria release some of this hydrogen as a gas. Algae and other plants also release oxygen. The amount of oxygen they return to the air is far greater than the amount of hydrogen. Almost all of the hydrogen that plants produce is consumed in making the special chemicals that lead to the formation of hydrocarbons and oxygen. Another aim of artificial photosynthesis, therefore, is to increase the amount of free hydrogen released in the water-splitting, carbohydrate-making photosynthesis process. Geneticists and biologists are also at work trying to increase the yield of hydrogen produced by living algae, but so far without great success. No hydrogen is released unless sunlight is available, just as Ingen-Housz discovered was the case for oxygen.

After free hydrogen in the form of hydrogen ions is generated by the absorption of sunlight by chlorophyll to produce an active hydrogen-containing enzyme, another complex cascade of chemical reactions occurs. This second series of chemical reactions doesn't need sunlight. It results in an enzyme whose chemical name (ribulose-1,5 carboxylase/oxygenase) is usually shortened to RuBisCo. RuBisCo completes the photosynthesis process by combining carbon dioxide molecules with hydrogen to form carbohydrates, such as corn sugar (glucose and dextrose), fruit sugar (fructose and levulose), and many others. One of these other carbohydrates is cellulose, a polymerized form of sugar.

The complex chemical reaction in going from water, sunlight, and carbon dioxide to carbohydrates and oxygen involves a bewildering variety of different structures in different plants. One result is the production of complex hydrocarbons, especially sugars of various types, as well as cellulose and oxygen.

Humans cannot digest polymerized sugar because our stomachs lack the enzyme that breaks cellulose apart. That's why we can't get nourishment from grass as cows and sheep can. As discussed in Chapter 12, making ethyl alcohol from corn starch and sugar is easy, but making it from polymerized sugar, like that in the cellulose in switchgrass or the cellulose-based wood produced by trees, is more difficult. Switchgrass could be four times more energy effective in producing ethyl alcohol (ethanol) than corn. Processing cellulose into alcohol requires multiple enzymes to break the cellulose apart so that the bacteria that produce alcohol can eat it. It is also possible, however, to thermally decompose the so-called lignocellulose that makes up most of the substance of wood to produce methyl alcohol, sometimes called wood alcohol. This thermal decomposition of wood to produce methyl alcohol can also produce tar, a product very important in the early days of wooden sailing ships, which needed tar to reduce the rate at which they leaked. North Carolina is called the Tar-Heel state because in its early days it was a major source of tar for the British navy, and was produced from the combustion of trees with insufficient oxygen to burn them completely. Methyl alcohol is also produced from wood in a similar fashion. Unlike wood, plain cellulose does not turn into alcohol or tar via thermal decomposition. Incidentally, unlike ethyl alcohol, which humans can and do drink in a variety of concoctions, methyl alcohol is poisonous.

Artificial photosynthesis might offer a means of producing sugar directly from sunlight, water, and carbon dioxide, after which making ethyl alcohol from that sugar to power cars involves only fermentation, brewing, and distillation. Artificial photosynthesis might even produce ethyl alcohol directly, without the need for fermentation. But the chemistry needed to produce either alcohol or carbohydrates from carbon dioxide and water isn't simple. It also may be possible to genetically engineer bacteria that can eat cellulose directly and produce ethyl alcohol without the need to depolymerize cellulose, but so far no one has been able to do this. Artificial photosynthesis appears to be one promising approach toward increasing the fraction of the energy in sunlight that can be turned into fuel. Water splitting is another.

Artificial Photosynthesis and Water Splitting

Ultimately, the goal of artificial photosynthesis is to harness as much of the energy in the solar spectrum as possible to produce useful chemicals. Nature has provided proof that some of the solar spectrum can be used by self-assembling organisms for this purpose. Plant chemistry provides part of a chemical blueprint for the production of fuels in an environmentally clean and renewable way. Achieving this result *in vitro* is not as simple as producing chlorophyll in quantity, because chlorophyll extracted from plants is not stable in sunlight outside of them. The reasons for this instability are not fully understood. Because isolated chlorophyll is not stable, attempts have been made to produce and use other stable, light-absorbing compounds instead. One class of materials that shows promise is ruthenium trisbipyridines, or Rubipy for short. Rubipy has been used in combination with exotic layered materials, such as potassium hexaniobate, to produce hydrogen via transfer of the electrons produced by light absorption in the Rubipy molecule to hydrogen ions in water. The structure needed to accomplish this feat is complicated and consists of layers of potassium hexaniobate separated by layers of Rubipy. So far, this process produces hydrogen at a far lower efficiency than simply using solid-state solar cells to electrolyze water directly.

The process using potassium hexaniobate consists of water splitting, without carbohydrate production. It therefore mimics only the water-splitting portion of photosynthesis. Splitting water directly into hydrogen and oxygen is more efficient than using the electricity from solar cells to electrolyze water. Most importantly, this process also might be cheaper than using solar cells and electrolysis, unless very low-cost, thin-film solar cells can be developed. For example, as the temperature rises, the fraction of the energy in sunlight that can split water into hydrogen and oxygen increases because of the tendency of the bonds in water (or other substances) to fly apart all by themselves at high enough temperatures. Living systems have severe limits to the elevated temperatures they can withstand. Artificial water-splitting techniques might be able to use far higher temperatures than living systems can. Higher-temperature chemical processes could produce hydrogen and oxygen directly from water at rates unmatchable by any living system. Electrochemists have succeeded in improving the efficiency with which water can be broken into hydrogen and oxygen by making use of electrolysis enhanced by sunlight.

Besides hydrogen and oxygen, direct production of other chemical species, such as hydrogen peroxide (H_2O_2), has been achieved by photoelectrochemical synthesis involving light as well as electricity. Hydrogen peroxide, a very energetic chemical used in concentrated form as an oxidizer for rocket fuel and in diluted form as an antiseptic, has been produced in this way. The number and variety of chemical species that could be produced by artificial photosynthetic or photoelectrochemical processes is almost boundless, and that is a problem: where to start? It will take an extended research effort by many different groups working a diverse assembly of approaches to determine which processes offer real possibilities for producing useful chemicals in large quantities with the help of sunlight.

So far, progress has been modest. Time will tell whether any of the numerous approaches to the generation of useful chemicals with the help of sunlight is as practical as simply using the photosynthetic chemistry that already works in green plants. Meanwhile, there are many other things that might be done to mitigate global warming, as we will see in the next chapter.

Recommended Reading

Ingen-Housz, John. *Experiments Upon Vegetables, Discovering Their Great Power of Purifying the Common Air in the Sunshine, and of Injuring it in the Shade at Night, to which is joined, A new Method of examining the accurate Degree of Salubrity of the Atmosphere*, London, 1779. Excerpts are available at http://web.lemoyne.edu/~giunta/INGENHOUSZ.HTML

Lawlor, D. W. *Photosynthesis.* 3rd ed. New York: Oxford University Press, 2001.

20
Planetary Engineering and Terraforming

We are as much gainers by finding a new property in the old
earth as by finding a new planet.

Ralph Waldo Emerson, *Representative Men*, 1850

Humanity is in the process of altering Earth's climate, as well as many other aspects of our environment, so in a sense we are already engaged in uncontrolled planetary engineering. Purposeful, directed changing of our world might mitigate the climate effects we have already set in motion. Terraforming, on the other hand, means something quite different. This word was coined in a series of stories written during the war years 1942–1943 by the American science fiction author Jack Williamson, but the basic concept of modifying other planets to make them habitable is of earlier origin. British writer Olaf Stapledon, in his memorable book *Last and First Men* (1930), envisioned the mass electrolysis of what we now know to be the non-existent oceans of Venus to produce oxygen on a planetary scale. A bit earlier, British writer and geneticist J. B. S. Haldane (1892–1964) speculated on humanity's survival on a changed planet Venus if Earth were destroyed by a cosmic catastrophe.

The imaginative leap inherent in terraforming does not fade from mind easily, and numerous science fiction writers, as well as many others, have conceived schemes for making Mars habitable. As human civilization pushes forward and deals with the impending energy and climate crises we face, terraforming Mars might actually be possible, as we will see later on in this chapter. Of more immediate import is the role that engineering on a planetary scale might play in compensating for the rapid growth in Earth's atmospheric carbon dioxide. Currently, dealing with this carbon dioxide problem seems mostly to involve proposals such as taxing the discharge of carbon into the air, injecting carbon deeply into the ground or the oceans, and planting lots of trees, or at least slowing down the cutting of the ones we already have, as well as increasing the efficiency of energy use and developing sources of energy and power that do not involve burning fossil fuels. These approaches should all be tried, sooner rather than later, and most of them are underway. But planetary engineering is a strategic approach towards solving global warming that should not be ignored. Let's begin first with how to change the Earth's albedo.

Energy Demand and Climate Change: Issues and Resolutions. Franklin Hadley Cocks
© 2009 WILEY-VCH Verlag GmbH & Co. KGaA, Weinheim
ISBN: 978-3-527-32446-0

Changing Earth's Albedo: Atmospheric Aerosols

You will remember from Chapter 2 that albedo is the fraction of sunlight that is reflected back into space. Is it possible to reflect more sunlight and cool the Earth down? The short answer is yes. One way to do this is to create more white puffy clouds. Another is to generate what amounts to smog in the upper atmosphere. Doing either of these things on a planetary scale isn't easy. But the needed changes in albedo are not stupendous or impossibly large. Compensating for half of the warming effect of doubling atmospheric carbon dioxide requires as little as a one-half percent change in albedo. Creating more white clouds seems feasible if nature helps. Don't forget that white puffy clouds are made of tiny droplets of water and that global warming will increase the evaporation of water from the oceans as well as the land. So far, however, the increased rate of water vapor evaporation caused by global warming has not caused enough additional cloud cover to stifle further warming.

Cloud seeding was developed and used in the later 1940s to increase rainfall. It involves dispersing tiny particles of something like silver iodide, which is highly effective in increasing nucleation of water droplets in moist air. If you want to experience a related nucleation effect, try stirring a spoonful of sugar into a glass of champagne or beer and see what happens (be sure to do this experiment over the sink). The technique of artificially nucleating raindrops and increasing rainfall works, but one silver iodide particle is needed for each drop of water. You can imagine how much silver iodide would be needed to accomplish this on a planetary scale. No one seriously suggests such a thing. To begin with, the silver iodide particles are cleared from the air by the resulting rainfall. Instead, generating smog in the upper atmosphere seems to be far more feasible on a vast scale, even if the very concept is abhorrent to every astronomer on the planet.

Major components of smog are sulfur compounds of various kinds. There is a lot of sulfur around, and it's not expensive. Putting sulfur, in aerosol form, high into the stratosphere is the problem. If these sulfur aerosols are not projected to extreme altitudes, they soon come back to Earth in the form of acid rain.

Nature accomplishes the feat of throwing things to high altitude by occasionally blasting megatons of dust and sulfur compounds sky-high via volcanic eruptions. The resulting changes in albedo do indeed cause planetary cooling that can last for more than a year. To make this effect last longer, the particles of sulfur compounds must be very, very small, tinier than one micron (one millionth of a meter). Bigger particles fall back to Earth faster than smaller ones. Dust and smog due to industrial activity, especially burning coal and driving cars, already act to partially compensate for increasing atmospheric CO_2, but these pollutants at low altitude also make people ill and are cleared away by rain. Smog in the upper atmosphere lasts longer than smog in the lower atmosphere and doesn't make people sick. Producing smog at high altitude offers one messy, but potentially effective, way of mitigating global warming. It turns out that one gram of aerosols of optimized size at high altitude could mitigate for a year or two the warming effect of more than 100 kilograms of CO_2. Since aerosols at higher altitudes last the longest, what's the best way of getting them up there? Not rockets, which are far too expensive. Old 16-inch naval guns can fire 2200-pound (1000-kilogram) projectiles to altitudes of 12.3 miles (20 kilometers), and this process costs a lot less than using rockets. Mitigating the effect of doubling atmospheric carbon dioxide would require continuously firing one shot every six seconds, assuming the resulting aerosols last two years. Such an artillery effort is not easy or cheap, but it can be done. Planes also can be used to deliver aerosols to lower altitudes, but aerosols won't survive as long there as they do at higher altitudes. Even though lower-level aerosols last for relatively short times, only about 150 flights a day are needed, if each flight delivers 100 tons of aerosols. Compared with the thousands of passenger planes that take off and land each day, 150 extra flights is not a large number.

Another approach is to use balloons to deliver dust and sulfur to very high altitudes. One problem here is the fact that as the altitude increases the air becomes thinner and the lifting power of a balloon decreases. To reach very high altitudes (20 miles or 32 kilometers), extremely large balloons are necessary, because the balloon's lifting ability increases as the cube of its size, while the weight of the fabric of the balloon itself increases only as the square of the balloon's size. This is why passenger-carrying zeppelin airships like the *Hindenburg* were so enormous. Huge balloons, perhaps 1000 feet (305 meters) in diameter, have been flown to extreme altitudes and may or may not have been the cause of some flying saucer reports. Filling giant balloons with helium is expensive, but filling them with hydrogen costs less. Other things being equal, a hydrogen balloon can lift about twice as much as a helium balloon. But helium does not burn, and hydrogen certainly does. If a giant hydrogen balloon explodes at low altitude, it will very likely kill somebody. Also, balloons inevitably return to Earth, and their remnants need to be collected. Tethering a high-altitude balloon dramatically reduces payload capacity because of the weight of the tether. At the moment, the best approach seems to be the use of large guns to launch sulfur aerosols high into the atmosphere, but their associated noise would limit where they could be fired. Even so, rough as these considerations are, they seem to indicate that changing planetary albedo is possible. But it is always dangerous to fool around with Mother Nature.

Tinkering with Planet Earth

Planetary engineering of the Earth is tricky because we all live here, and making a mistake could be catastrophic. For example, if a major volcano decided to explode while an upper-atmosphere smog project was in full swing, the planetary albedo could change a lot more than desired. Also, unplanned chemical interactions in the upper atmosphere might occur between sulfur and everything else up there, including chlorofluorocarbons. Human activity and chlorofluorocarbons have already affected the ozone layer, which shields us somewhat from ultraviolet light. If this ozone layer were destroyed and the ultraviolet level of sunlight dramatically increased as a result, an epidemic of skin cancer could result. Hydro-chlorofluorocarbons are being increasingly substituted for chlorofluorocarbons because they contain less chlorine and do less damage to the ozone layer. There might be other, but unanticipated, ill effects from filling the stratosphere with thousands of tons of sulfur aerosols. Caution is the watchword, and it turns out that there are other ways to decrease sunlight than making smog, as we will see next.

Parasols, Artificial Sunspots, Space Mirrors, Solar Sails, and Space Dust

The idea of blocking sunlight before it reaches the Earth is simple in concept and complicated in practice. Blocking one percent of sunlight would cancel out the global warming effect of doubling atmospheric CO_2, but how can we block out sunlight before it reaches the Earth? One way is by constructing what amounts to a very thin but gigantic parasol shading the Earth. On the downside, astronomers would be less than enthusiastic, and this parasol would take an awful lot of rocket launches to construct and cost trillions of dollars. Some have suggested that using electromagnetic launching would lower the cost a lot. For the moon, this might work (see Chapter 21). But for the Earth, electromagnetic launching is handicapped enormously by the very high speeds that must be reached at low elevations where the atmosphere is thick. Normal rockets start off slowly and only reach very high speeds at high altitudes, where the atmosphere is thin. Air friction increases as the square of the velocity, so extremely high speeds (e.g., 20,000 miles or 35,400 kilometers per hour) produce enormous air drag, violent sonic booms, and incredible air friction heating, all of which are not good. It might be possible, however, to use high-speed launching as a virtual first stage of a rocket to reduce the cost of getting things into orbit.

A virtual first-stage rocket boost from the controlled combustion of natural gas/air fuel in an inclined tube-launching assembly built along the western slope of a high mountain in the Pacific, perhaps on one of the many towering mountains on Hawaii, has been evaluated. High-velocity, jump-start launching of rockets of large size isn't easy. But in spite of high air resistance, jump-start launching dramatically increases the fraction of the mass of the original rocket that can be delivered as payload to orbit. The space shuttle delivers only 7.7% of the original vehicle mass to orbit. Analysis shows that a vehicle launch in jump-start mode might deliver 24.1% of the original vehicle mass to orbit. But humans could not be part of this mass because the needed accelerations during launching would be about 100 times the acceleration of gravity and would be lethal. Jump-start rockets can carry only cargo, but they might lower the cost of sending mass into orbit by more than a factor of three.

Using a single mirror to reflect one percent of sunlight might be unmanageable, even if it were placed between the Earth and the sun at the point where gravity and centrifugal effects are in balance (see Chapter 21). Such a reflector would appear as an artificial sunspot. The mirror also would be very hard to stabilize in place, for many astrophysical reasons, including counterbalancing the pressure of the sunlight. It would also have to be extremely large, since this balance point between the Earth and the sun is about a million miles (1.6 million kilometers) from Earth. Deploying many smaller mirrors closer to Earth in the form of large solar sails kept in orbit by continuously maneuvering their inclination so that the faint pressure of sunlight doesn't slowly waft them away is not impossible, just very difficult. Dispersing space dust between the Earth and the sun would also theoretically work, remembering, however, that dust particles smaller than a micron, about the size of a human red blood cell, would be blown clean out of the solar system by the faint pressure of sunlight. There are numerous problems with all of these space concepts, not the least of which is their stupendous cost. It would be far easier to change the Earth's albedo by making the surface of our planet reflect more sunlight, just as glaciers do. As it happens, many glaciers are melting, and their reflective effect is decreasing, not increasing. So what else can be done?

White Roads, Reflecting Roofs, and Shiny Balloons

Remember that the goal is to make the Earth reflect more sunlight. Whitewashing roads and using reflecting shingles for all roofs would be steps in the right direction. The trouble is that the effect of such steps is very small. Enormous numbers of large aluminized balloons at very high altitude might have more than a minimal effect. Reflecting balloons in orbit are also a possibility. In the days before transponder satellites could send signals across the Atlantic, the largest satellite ever produced was the Echo 1A satellite launched on 12 August 1960. The Echo balloon satellite was 100 feet (30.5 meters) in diameter and constructed from aluminized plastic. It was used to reflect radio signals between England and the United States. Brighter than most stars, this satellite may have been seen by more people than any other space object ever launched. Because of its low size-to-mass ratio, the pressure of sunlight and the solar wind pushed it out of orbit to destruction in the atmosphere on 24 May 1968. Releasing billions of reflective floats into the ocean rather than launching them into space is also possible but impractical. One possible exception: if the loss of ice cover in the Arctic were discovered to lead to catastrophically increased snowfall in the Northern Hemisphere, then covering Arctic seas with reflecting floats might suddenly seem more practical. In the end, one of the most achievable possibilities for cooling the Earth off by planetary engineering is producing more clouds over the open waters of the Pacific Ocean.

Back to Clouds Again

If not at extreme altitude, the smog that sulfur compounds produce can form tiny droplets of water and turn themselves into stratocumulus clouds without any silver iodide. The resulting drizzle of slightly acidic rain is how the lower atmosphere cleans itself up smog-wise. Rather than try to do this over land, where people live, the best place is over the open ocean. Producing a 4% increase in oceanic strato-cumulus clouds might require about 30% more cloud condensation nuclei than normally exist there. Stratocumulus clouds typically form above 50,000 feet (15,000 meters), and the resulting drizzle they can produce removes the nuclei that form them, so these nuclei continuously have to be replaced. After a lot of calculations, it seems that it would take the continuous addition of 32,000 tons of sulfur dioxide each day to increase oceanic cloud cover by 4%. This amounts to the sulfurous exhaust from 365 large coal-fired power plants. The net effect of a 4% increase in stratocumulus clouds would compensate for 300 parts per million by volume (ppmv) of added carbon dioxide, which is a lot considering that the present atmo-spheric level of carbon dioxide is about 380 ppmv.

Operating the ships necessary to inject this much sulfur dioxide into the air above the open oceans would cost a few billion dollars, but not the thousands of billions needed to build parasols in space. Ships that burned sulfur directly could be used, but sulfur also could be added to the fuel burned by already-existing cargo ships (with provisions for turning this extra sulfur off when they approach land, of course). As the price of oil increases, it might prove economic to go back to using coal to power cargo ships, and burning coal inevitably produces sulfur com-pounds. Powering cargo ships with coal, not oil, could lower fuel costs at the same time that it could increase the production of sulfur compounds over the water. Of course, burning coal also produces carbon dioxide, so using very high-sulfur coal is important. Spiking coal with sulfur, and maybe some iron powder as well (see the next section), also would help.

At the present time, the main addition of sulfur to the air over the oceans comes from marine algae, and the extra sulfur needed to increase cloud cover over the open ocean amounts to about 25% more than that from ocean algae. The increased acidity of the rain over the oceans might therefore be modest, but experts disagree strongly on this point. One controversy, for example, concerns the possibility that local effects could be much larger than global effects. Finally, as the temperature of the oceans increases due to global warming, the evaporation rate of water increases strongly, making more water available for cloud formation. Doing any-thing like this, however effective it might be, will require international cooperation on a scale that is seldom seen.

Feeding Algae

Marine algae in the open ocean, especially in Arctic and Antarctic waters, are starved for iron. It has been proven that feeding them iron causes marine algae to grow rapidly. Acid rain in the form of dilute sulfuric acid produced by burning sulfur-rich coal to fuel ocean cargo ships might increase production of airborne sulfur compounds from marine algae by making more sulfur available to them. And it would do this while increasing ocean cloud cover as we discussed in the preceding section. Also, powdered iron can burn, and adding powdered iron to sulfur-rich coal and using this mixture to power ships might increase both algae growth and clouds. One problem is that when masses of algae die, they may not sink to the bottom but rather decompose near the surface, releasing carbon dioxide or, worse yet, methane into the air. One possibility would be to harvest this ocean algae and then turn it into biofuel, whether biodiesel or alcohol. In that way, it could be used to power cars before its carbon was returned to the atmosphere as carbon dioxide

Inducing rapid growth of algae on a grand scale in the ocean is bound to cause unexpected problems of many kinds, not the least of which might be dramatic changes in the distribution and relative number of the various creatures that feed upon algae, phytoplankton, and seaweed. All of these factors and many more need to be studied thoroughly and this planetary engineering experiment tried on modest scales before mass implementation is attempted. But at least these options offer a chance to do something to mitigate global warming without decreasing economic activity. Beyond these planetary engineering concepts for the Earth, making at least one of the sun's other plants habitable is a dream, but not an impossible one. Since nobody seems to live on any of these planets, the possibility of disaster is greatly reduced. Let's see how such things might work.

Terraforming Mars (and Maybe Venus)

Terraforming Mars has the advantage that no people live there, so tinkering with Mars is not a perilous as tinkering with Earth, where everybody lives. The American astronomer Percival Lowell (1855–1916) devoted his fortune to looking for the "canali" that the Italian astronomer Giovanni Schiaparelli had reported in 1877. Mistranslating *canali*, which means "channels" in Italian, into the English word *canals*, Lowell imagined that these features represented a planetary engineering effort by Martians to bring water from melting Martian polar ice to the equatorial regions of their desiccated and dying world. Lowell's concept was supported by the seasonal changes that are observed in Martian surface features through even modest telescopes.

The scion of a wealthy Bostonian family, Lowell was able to follow up on his dream by building an observatory under the clear skies on the outskirts of Flagstaff, Arizona. His most famous book is *Mars and Its Canals* (1906). (He died and is buried near his observatory, in a mausoleum whose stained glass roof mimics the deep blue vault of the evening sky. It was at this observatory in 1930 that Clyde Tombaugh discovered the planet-like object he named Pluto, possibly because the first two letters of this name are Lowell's initials. We know now, of course, that no Martians made any such planetary engineering effort to save their dying world. Nonetheless, there is indeed moisture on Mars, trapped as polar ice mixed with solid carbon dioxide and adsorbed onto the surface of Martian dust. There may even be some liquid water buried deeply beneath the Martian surface. Numerous photographs give compelling evidence that there once was lots of water on Mars, leaving traces of its existence in numerous channels, none of which was produced by Martians. The ancient oceans on Mars are long gone, in part because of the weak Martian gravity. The thin atmosphere of Mars also lets a disproportionate amount of short-wavelength ultraviolet light through to the surface, and this light can break the bonds of water molecules. On Mars, the surface gravity is only about 40% of ours, and the atmospheric pressure is only about 0.7% that on Earth, and 95% of it is carbon dioxide. The rest is mostly nitrogen and argon, with only 0.03% water vapor. Only a limited amount of water still remains on Mars, while oceans of it remain on Earth. Many concepts for terraforming Mars involve returning vast amounts of water to its atmosphere. How could such a feat be accomplished?

It is likely that all the water on Earth and Mars came originally from the clouds of ice and comets enshrouding the solar system. Before Jupiter "vacuum cleaned" most of these objects from the inner solar system, impacting "dirty snowball" comets were very likely a common event in Earth's distant past. Comet impacts are no longer common events, but icy impacts still occur from time to time. The famous Tunguska impact event in Siberia in 1910 may have been one of them. The Tunguska event was the largest Earth–object impact in recorded history, although, interestingly, it produced no crater, just a giant explosion that probably occurred above ground level. It's still not clear what kind of object it was that hit us in 1910, but an icy mass is certainly one possibility.

Jupiter is still cleaning up the inner solar system, and in 1994, to the delight of astronomers and the education of the rest of us, remnants of a comet were photographed impacting the surface of Jupiter. On Earth, the water from eons of icy comet strikes has remained to form our oceans. On Mars, most of the water from impacting comets is gone. One factor that helps Mars keep at least some water is its weak sunlight, outside the ultraviolet range, which makes the planet very cold. The sunlight on Mars is only about 43% as bright as ours, and the average temperature is −56 °C (−100.8 °F). Vaporizing the carbon dioxide and water vapor frozen on one or the other of the Martian poles, and keeping it vaporized, would cause Martian global warming as well as increasing its atmospheric pressure. It turns out, however, that the effect of vaporizing the ice and carbon dioxide at the Martian poles is not large enough to make a very big difference in the average Martian temperature, mainly because the Martian polar caps are mostly frozen water and frozen carbon dioxide, and water is not as effective per molecule in producing global warming as carbon dioxide is. Also, this water would just freeze out somewhere else on the surface if vaporized at the poles.

Something stronger is needed, such as chlorofluorocarbons or hydro-chlorofluorocarbons, whose greenhouse effect can be thousands of times greater than that of water and hundreds of times greater than carbon dioxide. And they wouldn't freeze out of the Martian air as easily as water does. Even so, an awful lot of chlorofluorocarbon gas would be required to warm Mars up enough to just be very cold, and chlorofluorocarbons are not cheap. Using enormous amounts of water, not lots of chlorofluorocarbons, might be more practical, even if water is not as effective a global warming gas as chlorofluorocarbons. Even when ice produces water vapor, the vapor pressure of ice at its freezing point is just a little less than the atmospheric pressure on Mars. It turns out that water is a very common substance in our universe, being composed as it is of hydrogen, the most common element there is, together with oxygen, which also exists in great quantity. Most of the water in our stellar neighborhood is locked up in the form of innumerable icy objects enshrouding the outskirts of the solar system. There is more water on the outskirts of the solar system than anyone could possibly want. Getting this water to Mars is the problem, and it is one that would take tremendous energy and effort to solve. Solar sails might serve to gradually drag icy masses from deep space to Mars, but this method would be very slow. Another way of increasing the temperature on Mars would be to decrease its planetary albedo, which is already less than half of Earth's, by making the surface of Mars absorb more sunlight than it does already. For the Earth, the problem is an opposite one: how to increase Earth's albedo and reflect more sunlight to counteract global warming, as we discussed earlier in this chapter.

Terraforming Mars wouldn't be fast or easy, but it's not impossible. Terraforming Venus would be far more difficult. To begin with, global warming has run amuck on Venus. Its dense atmosphere of carbon dioxide is about 100 times as heavy as Earth's atmosphere. This massive carbon dioxide atmosphere, together with the proximity of Venus to the sun, gives it a surface temperature of up to 900°F (482°C). It even rains hot sulfuric acid there. What could be worse? Changing these extremely nasty conditions enough to make it livable would be a truly colossal task.

It is unfortunate that there doesn't seem to be any way to easily move some of the carbon dioxide on Venus to Mars. If there were, we could kill two birds with one stone, so to speak, and terraform both planets at the same time, decreasing global warming on Venus while increasing it on Mars. Unfortunately, doing so is way beyond current human ability.

About the only approach that seems even remotely feasible is creating some kind of extremely hardy bacteria that turns carbon dioxide into oxygen plus carbon. As it happens, an Earth-bound variety of red algae that lives in hot acidic springs can actually thrive on carbon dioxide. But it can't survive the high temperatures on the surface of Venus. No known life form can. So what can possibly be done? For one thing, the upper atmosphere of Venus may be cool enough for red algae to survive, and as red algae in the upper atmosphere ate carbon dioxide, the planet would cool down as the carbon dioxide was slowly consumed. The red algae might survive at lower and lower altitudes, until finally reaching the surface. Any such planetary engineering would require millennia, if not longer. In the end, terraforming either Mars or Venus must wait until humanity solves its current energy and climate predicament. If we don't solve it, then the question of terraforming other planets in our solar system becomes moot.

What Can Be Done?

As things stand now, generating increased cloud cover over the open oceans appears to be one approach to planetary engineering that is practical enough to be seriously attempted relatively soon. Trying this on a modest scale is an achievable goal that also might show what could go wrong. It may be that many expensive world conferences involving thousands of delegates and a large number of plane trips to exotic locations would have to occur before any such attempt is made. Or, if the will is there, perhaps low-cost teleconferencing would suffice, at modest financial and energy cost. Teleconferencing such meetings would set a good example of what can be done to decrease energy use as we struggle to fix Earth's energy and climate problem. If we don't solve this problem somehow, things could become very bad indeed, as we discuss in Part IV. But there is an additional approach towards supplying humanity with nearly unlimited energy that we haven't considered yet. This method involves generating electric power from the massive amount of solar energy available in space. Let's consider that in the next chapter.

Recommended Reading

Fogg, Martyn J. *Terraforming: Engineering Planetary Environments.* Warrendale, PA: Society of Automotive Engineers, 1995.

National Academy of Sciences, National Academy of Engineering, National Institute of Medicine. *Policy Implications of Greenhouse Warming: Mitigation, Adaptation, and Science Base.* Washington, D.C.: National Academy Press, 1992.

Sagan, Carl. *The Cosmic Connection: An Extraterrestrial Perspective.* Garden City, NJ: Anchor, 1973.

21
Space Solar Power: Energy and the Final Frontier

Throughout space there is energy. If static our hopes are in
vain; if kinetic – and this we know it is for certain – then it is a
mere question of time when men will succeed in attaching their
machinery to the very wheel work of nature.

Nikola Tesla, Quoted in *The New York Times*, September 30,
1894

Nikola Tesla (1856–1943), the Croatian-American inventor of the alternating-current electric motor and numerous other electrical devices, was probably thinking about electromagnetic energy and not the kinetic energy of planets spinning around in orbit about the sun when he was quoted in *The New York Times*. Nevertheless, there is indeed a colossal amount of energy in the planetary "wheel work of nature," but harnessing it is beyond current human ability. On the other hand, harnessing the power in sunlight is well within our reach, as we have seen in Chapter 6. Since the sun always shines in space, it seems to be the perfect place for solar cells. But there are problems. For one thing, it costs a lot to get there. For another, the level of ionizing radiation in space is high enough to damage solar cells unless a transparent protective layer is used to shield them. Glass turns dark when it is irradiated, so polished sapphire cover plates are used instead. That increases cost. Also, things in orbit move around.

Even so, the total power output of the sun is enough to supply all human energy needs billions of times over, if only some way could be contrived to harness it. One big problem is that by the time sunlight reaches Earth's orbit, it is spread out a lot and its level of power per unit area is reduced. Gathering in lots of solar power requires very big collectors, and huge arrays of solar panels are costly. Besides cost, another major problem is where to locate these solar panels in space. A third problem is how to get the electrical power produced in space down to Earth. Let's look first at where to put solar power panels and solar satellites.

Lagrange and His Famous Points

The French-Italian mathematician Count Joseph-Louis Lagrange (1736–1813) possessed one of the greatest mathematical intellects of his era. His special field was

Energy Demand and Climate Change: Issues and Resolutions. Franklin Hadley Cocks
© 2009 WILEY-VCH Verlag GmbH & Co. KGaA, Weinheim
ISBN: 978-3-527-32446-0

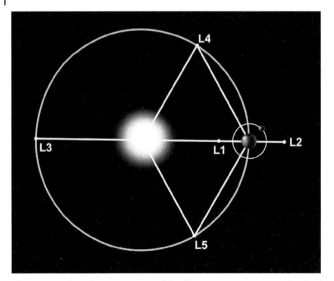

Figure 21.1 The relative position of the five Lagrange points in the Earth–sun system.

celestial mechanics, and his first major prize was awarded by the Paris Academy of Sciences for his explanation of the moon's oscillations, those small shifts that cause slight changes in the portion of the edge of the moon's surface visible from Earth. Lagrange was invited to Berlin by Frederick the Great of Prussia, who is reported to have declared "the greatest king in Europe" wished "the greatest mathematician in Europe" to attend his court. It was Lagrange who first pointed out that in two-body systems, such as the sun and the Earth or the sun and Jupiter, positions can exist where small bodies can remain indefinitely. There are five Lagrange points in the Earth–sun system, as shown in Figure 21.1.

Of these five, only two (L4 and L5) are stable, meaning that small bodies will remain there even when slightly disturbed. The other three (L1, L2, and L3) are unstable, and slight disturbances can displace objects from them. Unfortunately, as Figure 21.1 shows, the two stable points are located along Earth's orbital path, one-sixth of the way around the sun and many millions of miles in front of the Earth (L4) or behind (L5) it. In 1906 the stable Lagrange points in the Jupiter–sun system were found to contain asteroids, the so-called Trojan satellites. In the Earth–sun system, the L2 point is more than twice the distance of the moon on the night side of the Earth. The L3 point is on the opposite side of the sun from the Earth. For space solar power, the important Lagrange point is L1. Although not one of the stable points, it is on a direct line between the Earth and the sun, about one million miles (1.6 million kilometers) from Earth. At this point, a large solar parasol could function as a permanent sunspot, reducing the sun's warming power on the Earth while at the same time acting as a solar power station.

The pressure of sunlight on a parasol satellite at this L1 location adds difficulty to the engineering of such a sunlight shield. But sunlight pressure problems might be addressed by moving the solar power station slightly towards the sun from the

LI position, thereby increasing the gravitational attraction of the sun enough to overcome the pressure of light against the parasol. This displaced orbital position would then require some means of thrust for stabilizing it relative to the Earth. Slight tilting of the parasol offers another possibility, by changing the angle at which the sunlight is reflected from the parasol. Such an approach would not require a supply of rocket fuel. If this parasol were gigantic, something like 745 miles (1200 kilometers) in size, it would block about 2% of the sunlight that normally reaches the Earth and effectively counterbalance current global warming. The effect of such an artificial sunspot would have many times the effect of regular sunspots and would not come and go in an 11-year cycle. If this parasol were made from 10% efficient solar cells, the resulting electric power would be more than that from 100,000 normal power plants, far more than the electric power needs of our entire planet many times over. Getting electricity from such a parasol to the Earth would be another true engineering feat, involving microwave power transmission on a stupendous scale. This whole project is far beyond gigantic and would cost more than all the other possibilities described in this book combined. It is unlikely that anything like it will be attempted anytime soon. But at least this possibility exists, unbelievably difficult as it is. Closer to home there is another, more feasible possibility in the form of solar power satellites in geosynchronous orbits, so let's consider them next.

Geosynchronous Orbits and Solar Sails

Geosynchronous orbits are used for satellites that beam television programs to Earth. Objects in such orbits are fixed in position with respect to a given position on Earth's surface. In a geosynchronous orbit, a satellite is orbiting at the same angular velocity as the Earth. That is, such satellites rotate around the Earth once per day, just as the Earth rotates around its axis once per day. This fact means that these satellites appear to stay fixed in position in the sky overhead. That's why satellite television antennas don't need to be repositioned all the time.

Equatorial geosynchronous orbits are about 22,000 miles (35,400 kilometers) above the Earth. Only in equatorial orbits around the equator are satellites in such orbits naturally stable in a fixed position above the Earth. The basic idea is that in equatorial synchronous orbits, not only does the angular velocity of the satellite match that of the Earth but also the force of gravity exactly balances the centrifugal force needed to stay in orbit. For non-equatorial geosynchronous orbits, centrifugal force plus another force is needed to keep a satellite both in orbit and fixed above a constant spot on the Earth. This extra force can be supplied by the proper tilting of a solar sail. This result is important, since at the moment, there are so many geosynchronous satellites already in orbit around the equator that there is a shortage of positions for additional ones. Satellites move around a little due to solar wind and therefore need to be spaced far apart to avoid the chance of crashing into each other. Dr. Robert L. Forward (1932–2002) was the first to point out that the use of solar sails would enable geosynchronous satellites to be stable in non-equatorial geosynchronous orbits. Forward's solar sail concept makes thousands of additional geosynchronous orbital positions possible. Better still, making solar sails using thin-film solar cells kills two birds with one stone, because these cells would generate solar power while at the same time comprising the sail that stabilizes the non-equatorial geosynchronous solar power satellite above a fixed spot on the Earth.

Geosynchronous satellites have the advantage of being located a lot closer to us than the L1 Lagrange point is. Many geosynchronous solar power satellites would be needed to supply Earth's ever-growing appetite for electrical power. But there is plenty of space in solar sail–stabilized orbits to place large numbers of solar power satellites in the non-equatorial geosynchronous orbits envisioned by Robert Forward. Better yet, in such non-equatorial geosynchronous orbits, a solar sail is less troubled by the Earth's shadow, because in such orbits they cut through smaller segments of it. How can we get the power from the solar cells of these geosynchronous satellites down to Earth? One workable solution may be wireless transmission of power.

Beamed-power Microwave Transmission

Wireless transmission of power can be done in several different ways. Lasers, for example, can transmit power without wires. Beamed power is the general term for sending energy from one place to another using electromagnetic radiation, of which laser light is just one example. With current technology, the most practical way of sending massive amounts of power from space to Earth is by means of microwaves. Microwave electromagnetic radiation includes the frequency range normally thought of as radar. The length of such waves is much longer than that of light waves and much shorter than that of radio waves. The fact that microwaves are used for radar is an important reason why microwave technology has been developed so extensively, most notably in the 2.45-gigahertz (GHz) frequency range. Such frequencies are used in microwave ovens, as well as in many military and commercial radar installations. At 2.5 GHz, water absorbs microwave energy strongly. That's why this frequency heats food in microwave ovens so readily.

Far less atmospheric absorption occurs when higher frequencies, such as 35 GHz, are used, and a higher frequency can use a smaller antenna for reception. But the technology for higher frequencies is not as advanced as that for 2.45 GHz. A small helicopter was flown using 2.45 GHz microwave power more than 40 years ago, and a variety of small planes and helicopters have repeated this feat since then. In 1975, 30 kilowatts of power were transmitted over a distance of 1 mile (1.6 km) by means of microwaves. Of course, transmitting power via microwaves requires both a transmitting and a receiving antenna, and naturally there are energy losses. However, 54% overall efficiency of power transmission by microwaves has been demonstrated, and transmitting power by microwaves at an efficacy of 70% appears possible.

Beaming enormous amounts of power back to Earth using microwaves naturally presents a lot of problems, such as the dangerously high power level of intense energy beams, which could be hazardous to birds and planes flying through them. They also might be used as weapons of mass destruction. But from the technical point of view, the process appears feasible. Transmitting the equivalent of the power produced by five typical 1000-MW coal-burning power plants from space to Earth would require a transmitting antenna about 0.62 miles (1 kilometer) in diameter and a receiving antenna in the form of an ellipse about 6.2 by 2.1 miles (10 km by 13 km), if it is positioned at a latitude of, for example, 40 degrees. At higher latitudes than this, more of the beam is spread out on the Earth's surface, just as sunlight is, and a bigger receiving antenna is needed. In Robert Forward's solar-sail, non-geosynchronous orbits, the receiving antenna size at higher or lower latitudes is reduced because the angle between the sending and the receiving antenna is decreased for solar-sail orbits at higher or lower latitudes. Even in regular geosynchronous orbits, if the receiving antenna could be tilted upward from the ground to intercept the beam directly, not at an angle, then its size could be reduced, but tilting a massively large antenna is not easy. At the equator, of course, such tilting would not be necessary for normal geosynchronous orbits, as the solar satellite would be directly overhead. Building large microwave receiving facilities would be very difficult and costly, but not impossible. Above everything else, the cost of putting solar cells into orbit by using rockets from Earth dominates the feasibility of the solar power satellite concept. Currently, there appear to be two possible approaches to the problem of the extremely high cost of getting things into space. These two approaches are either elevators to space from the Earth or electromagnetic launchers on the moon. Let's look at space elevators first.

Space Elevators

An elevator into space sounds ridiculous, but it isn't. Launching things into space using rockets as we currently do requires an enormous amount of energy. Carrying things into space with an elevator, not chemical rockets, would tremendously reduce the energy cost of reaching orbit. Fundamentally, the energy needed to lift one kilogram of mass (2.2 pounds) from the surface of the earth to geosynchronous orbit is not that much, amounting to the energy needed to operate 10 100-watt light bulbs for one day. This fact represents about a 10,000-to-1 reduction in energy compared with using rockets. How could such a dramatic reduction in the cost of putting things into space be achieved? One answer has been given many names, ranging from "beanstalk" to "orbital tower" to "skyhook" or "cosmic funicular." In 1895 the famous Russian visionary and schoolteacher Konstantin Tsiolkovsky (1857–1935) came up with the idea of a tower rising from the equator into space. Then in 1960 the Russian engineer Yuri Artsutanov proposed an elevator (which he termed a "heavenly funicular") from the equator of the Earth to a height of 21,750 miles (35,500 kilometers) to achieve the goal of reaching space without using rockets. Is either a space elevator or a gigantic tower possible at all?

For a tower, the answer probably is no, because the total mass of material required becomes utterly unreasonable. A space tower must be of gigantic size at its base and taper as it reaches higher elevations, just as mountains do. But for a space elevator, the answer is, yes, but just barely. And in terms of the total mass of materials required, the elevator idea wins hands down but requires cables of extraordinary strength. It should be mentioned that variations of the space elevator have been proposed since Artsutanov's day, including orbiting, spinning, gigantic tethers that would periodically swing down low from their mother ship in orbit, like the proverbial "sweet chariot," to pick up people and cargo and carry them into space after picking them up and transferring them to orbit from a high-altitude, high-speed transfer vehicle. But even such imaginative rotating tether concepts need something much stronger and lighter than steel. The strongest steel wire ever made was produced for wire-guided missiles. But even this wire, with an ultimate tensile strength of about 500,000 pounds per square inch (3447 megapascals) breaks under its own weight after reaching a height of only 27.8 miles (45 kilometers), even though the force of gravity diminishes with elevation above the Earth's surface. Such heights are nowhere near the 22,000 miles (35,400 kilometers) needed to reach geosynchronous orbit. Of course, at a height of 27.8 miles a second wire could be added to the first, to increase the height that could be reached without the whole thing breaking, at which point a third wire could be added, and so forth, on up to 22,000 miles.

By continuing to add more and more wires to the cable, it is theoretically possible to build such a skyhook right now, but a little math shows that the number of steel wires needed rises exponentially, and the resulting cable grows rapidly to a ridiculous size and a total weight equivalent to thousands of battleships. Unlike a space tower, the cable of a space elevator starts small in size and increases as it reaches higher altitudes, at least with the strongest materials currently available. Steel wire is not the way to go. Steel wire is strong enough to lower things into the Marianas Trench, the deepest part of the Pacific Ocean, where the water extends about 7 miles (11 kilometers) down, but steel wire is not strong enough to reach anywhere near 22,000 miles into space. Reaching geosynchronous orbit requires something much stronger, and much less dense, than steel. What is the strongest possible light-weight material?

It turns out that the theoretical strength of solids increases as the square root of their modulus of elasticity, which is a measure of the force per unit area that it takes to stretch something elastically a given amount. A very hard material like a diamond has a modulus of elasticity that is five times that of steel, and better still, diamond is only about half as dense as steel, so it can reach to greater heights before breaking under its own weight. Even so, diamond wires are not much more practical than steel wires, since either a continuously increasing thickness of solid diamond or a tremendous number of diamond filaments connected together must be used. Something even stronger and lighter than diamond is needed, and that something is carbon nanotubes.

Nanotubes can be thought of as the carbon equivalent of nylon stockings. Nylon stockings can be easily stretched at first, but their modulus of elasticity increases as they are stretched further and the threads in the stocking begin to rotate toward the direction of the applied force. So it is with nanotubes of carbon. To better understand the structure of carbon in nanotubes, remember that graphite is a form of carbon. It is useful as a lubricant because the sheets of carbon from which graphite is made are very strongly bonded within the carbon sheets but only very weakly bonded from sheet to sheet. Graphite powder lubricant is used for car locks because it is dry and will not gum up locking mechanisms in very cold weather the way oil does. Nanotubes are effectively sheets of carbon graphite rolled up into tubes. They have a very high ultimate tensile stress, about 100 times higher than steel, because their carbon-to-carbon bonds can rotate toward the direction of the applied force, just as the threads in nylon stockings can when the stocking is stretched lengthwise, thereby increasing its modulus.

Carbon nanotubes are also much less dense than steel. Even so, a skyhook cable constructed using carbon nanotubes must still have a diameter that increases in size as the cable reaches higher altitude, if it has to extend from the surface of the Earth upward to 22,000 miles (35,400 kilometers) without breaking under its own weight. However, and this is the main point, it has been estimated that the simplest, smallest cable of carbon nanotubes capable of reaching from Earth to equatorial orbit would have a mass of only about 900 kilograms. Compare this to the thousands of battleships worth of steel it would take to build a space elevator from steel wire. Of course, no way is known of actually making such a cable, but the concept does not appear to be ridiculous, only extraordinarily difficult and, of course, costly. Building an elevator to space is not impossible, and if it could be built, Figure 21.2 shows what a space elevator might look like.

Once the cable has reached the geosynchronous orbit position, what holds it up? Why doesn't it just fall down? The answer is that it would indeed fall back to earth

Figure 21.2 A conceptual view of a space elevator as it might appear from the elevator itself.

if it just stopped at 22,000 miles (35,400 kilometers). But if it goes on beyond this point, then the centrifugal force produced by the rotation of the part of the cable that extends beyond 22,000 miles (35,400 kilometers) can hold the whole thing in place. The total length needed turns out to be about 40,000 miles (64,400 kilometers) unless an orbiting ballast mass is used, such as a captured asteroid tied to the end of the cable. The centrifugal force grows larger as the distance from the Earth and the length of the cable extends beyond 22,000 miles (35,400 kilometers). This means that an extended cable can act as a slingshot, similar to the one that David used to slay Goliath. It would be possible to launch payloads into space from the geosynchronous satellite point just by letting them slide further on up the cable.

Where does the energy come from to accomplish this launching feat? The answer is that it comes from the energy of the Earth's rotation, which would slow down infinitesimally with each launch. Infinitesimally is the important point here, since the slowing down of the Earth with each launch would be immeasurably small. By these slingshot means, humanity could actually harness one mechanical form of Tesla's "wheel-work of nature." There is such a stupendous quantity of energy in the rotating Earth that a solar system full of space colonies could be supplied for millennia without slowing the Earth down very much. Proposed in the 1970s by Princeton professor Gerard K. O'Neill (1927–1992), large colonies in space would provide a non-planetary habitat for humanity. Glorious as this all seems, there may not be time to seriously undertake such things before Earth's fossil fuels are exhausted. Still, it is a wonderful dream, and who knows, maybe some scientist or engineer can develop a material that is even stronger than carbon nanotubes and also easy to manufacture. Such a development would make the whole space elevator idea a real possibility, not just a dream. Meanwhile, it is more within the realm of known technology to build an electromagnetic launcher on the moon to supply the material needed to build gigantic solar panels in space from lunar, not terrestrial, materials.

Electromagnetic Launching

The moon has virtually no atmosphere, of course, and so it is possible to accelerate something to very high speeds using magnetic fields without worrying about air friction. On Earth, air friction is a severe handicap to moving anything very fast. The drag from air friction increases as the square of velocity. During the oil crisis of the 1970s, the maximum speed on highways in the United States was limited to 55 miles (89 kilometers) per hour. At this reduced speed, cars and trucks go further on a gallon or liter of fuel than they can at higher speeds, where the drag of air friction is much greater. On the moon, with no air, it is actually feasible to accelerate something to a high enough velocity that it can be shot directly into orbit even though it stops accelerating before leaving the lunar surface. On Earth, rocket ships continuously accelerate for several minutes and only reach orbital velocity after they have left most of Earth's atmosphere behind.

In 1950 the famed British science fiction writer Arthur C. Clarke (1917–2008) suggested launching payloads electromagnetically from the surface of the moon. Propelling objects to extreme velocity without using fuel, just electromagnets, provides a method for sending into orbit the great masses of material needed to construct O'Neill's space colonies and solar power satellites without using chemical fuels. The basic idea is to use sunlight and solar cells to power an electromagnetic launcher and deliver large masses of lunar materials into orbit for constructing space habitats and solar power satellites. Lunar soil can be processed into silicon to make solar cells, but it may actually be possible to use certain types of lunar minerals for solar cells directly. That's because some minerals are semiconductors, although not very good ones, and it might be possible to melt and form certain types of moon rocks into very low-efficiency solar cells without bothering to refine them into pure silicon. The mineral ilmenite ($FeTiO_3$), for example, exists on the moon, and it has semiconducting properties that in some ways resemble those of silicon. Ilmenite might, at least theoretically, produce 10% efficient solar cells as compared to the 25% cells that can be made from silicon. But undoubtedly, the actual efficiency of solar cells produced from lunar ilmenite would be far lower than 10%, for lots of semiconductor physics reasons.

Solar power satellites transmitting large amounts of electric power to Earth is only a dream. Not an impossible one to be sure, but a dream nonetheless. Humanity may be fast approaching that point where the resources left in our remaining fossil fuel energy supply are not sufficient to construct a large solar power satellite system while at the same time increasing living standards. The drain on the world's resources imposed by population growth, wars, increasing living standards, and the approaching end of plentiful oil, together with worldwide climate changes, might put the dream of solar power satellites permanently out of reach, for reasons discussed in Part IV.

Recommended Reading

Clarke, Arthur C. "Electromagnetic Launching as a Major Contributor to Space Flight." *Journal of the British Interplanetary Society*, 9, (1950): 261–267.

Glaser, Peter E., Frank P. Davidson, and Katinka Csigi, eds. *Solar Power Satellites: A Space Energy System for Earth*. New York: John Wiley, 1997.

Part IV Nightmares

Energy Demand and Climate Change: Issues and Resolutions. Franklin Hadley Cocks
© 2009 WILEY-VCH Verlag GmbH & Co. KGaA, Weinheim
ISBN: 978-3-527-32446-0

Introduction

Where are they?
Enrico Fermi, circa 1950

The renowned Italian-American physicist Enrico Fermi (1901–1954), discoverer of neutron-induced artificial radioactivity, directed the world's first controlled nuclear chain reaction on 2 December 1942, using a graphite-moderated reactor built under Stagg Field at the University of Chicago. This development marked the beginning of the nuclear age. After the possibility of nuclear Armageddon became obvious, he mused upon the survival of intelligent creatures in the universe and raised the question "Where are they?" This question is known as Fermi's paradox, because the enormous age of the universe, about 14 billion years, is so great that earlier intelligent life anywhere in our galaxy would have had ample time to extend completely across it, even at sub-light speeds, provided only that intelligent life can endure for long periods of time.

So far we have no real evidence that any intelligent life exists other than our own. Our galaxy comprises more than 100 billion stars and undoubtedly a vast number of planets, only a very few of which have been discovered. If advanced civilizations have existed elsewhere in our galaxy, what happened to them? There are many possible answers to Fermi's paradox. Catastrophic astronomical events can destroy dominant life forms of any kind, not just dinosaurs. Our own history shows that wars and the exhaustion of resources can destroy civilizations. The proverbial four horsemen of war, famine, disease, and death may ride on other worlds besides our own. If any civilization fails to solve its major problems, it cannot endure over long periods of time. On the other hand, if we can solve our approaching energy and climate crisis, then there is no limit to what humanity might accomplish.

Energy Demand and Climate Change: Issues and Resolutions. Franklin Hadley Cocks
© 2009 WILEY-VCH Verlag GmbH & Co. KGaA, Weinheim
ISBN: 978-3-527-32446-0

22
Alternative Futures

Limitless possibilities opened before the human imagination.

H. G. Wells, *The Shape of Things to Come*, 1933

... of all sad words of tongue or pen,
the saddest are these: "It might have been."

John Greenleaf Whittier, *Maud Muller*, 1854

This chapter is headed by two quotations for a good reason. Humanity has achieved the ability to do magnificent things, but it is far from certain that our great ability will be used wisely. It is difficult to predict which quote will be more applicable a millennium from now. As we have seen, there are many possibilities for supplying the energy and power that humanity needs, without irrevocably altering the climate of the Earth. But if humanity continues along the path it is now on, we will burn all the fossil fuels we can get our hands on. What happens then?

The ultimate limit to the use of fossil fuels comes when the energy cost of harvesting them exceeds the energy they produce when burned. The CO_2 generated by burning *all* fossil fuels on Earth (not just all reserves but also all resources) could increase the atmospheric level of this gas to over 12,000 ppmv, nearly 32 times what it is now. But this level may never be reached. Some fossil fuel reserves may be unharvestable. Climate changes may become so great that humanity would experience a great die-off, as many other species have done before us. Extremely large increases in atmospheric carbon dioxide may be limited because the oceans and the land, as well as the ecosystem, all interact with carbon dioxide. Remember that CO_2 is more soluble in cold water than in warm water. One horrific possibility is the dramatic increase in global temperatures that would be brought on if a portion of the enormous quantity of CO_2 in ocean water is released as the oceans warm up. Can humanity really cause such a monumental climate disaster?

Applying the carbon cycle 50% rule discussed in Chapter 4 to estimate how much carbon dioxide can endure for millennia in the atmosphere means that the level of CO_2 in the air would only reach 6000 ppmv, one-half the total delivered into the atmosphere from burning all possible fossil fuels. But even this reduced carbon dioxide level is still about 16 times higher than the level we have now. The global warming that could result from such a large increase in atmospheric

Energy Demand and Climate Change: Issues and Resolutions. Franklin Hadley Cocks
© 2009 WILEY-VCH Verlag GmbH & Co. KGaA, Weinheim
ISBN: 978-3-527-32446-0

carbon dioxide would be tremendous, amounting to an increase in the average temperature of the Earth of perhaps 27 °F (15 °C). It is notable that in the early days of our planet, when the sun was young and generated less energy than it does today, much of the carbon now stored in fossil fuels very likely existed as a high level of atmospheric carbon dioxide. In those long-ago days it may have been the global warming caused by this atmospheric CO_2 that kept our planet warm enough to help life begin. Over geologic time, the sun's output of energy has grown considerably greater than that of its youth, the CO_2 level has gone down, and we've grown used to the climate we have had for the last 10,000 years. It will take the slowly increasing energy output of the sun billions of years to heat the Earth up this much by itself. It may take us less than 500 years.

This balance between increasing output of the sun and decreasing atmospheric carbon dioxide was a good thing for the beginning of life on Earth, and it would also be a good thing if we could maintain this equilibrium. But the warming trend now in progress from increasing atmospheric carbon dioxide, together with our sun's present level of energy output, could lead within a few centuries to global atmospheric temperatures unparalleled in the millions of years that any creatures resembling humans have existed.

An enormous disruption in the climate conditions on the Earth would surely follow. The effect of such changes on civilization can only be imagined. Without proper action taken in good time, it might require only 100–500 years to reach a truly catastrophic climate situation, if we continue along as we now are, driven by the world's growing population and desire for increasing standards of living. After all possible fossil fuels have been burned, natural atmospheric, oceanic, and mineral processes might begin to slowly lower the level of carbon dioxide in the atmosphere. By then, the temperature of the ocean will also have risen substantially. Increasing ocean water temperatures will affect the speed and extent to which the oceans absorb CO_2 from the atmosphere. In a worst-case scenario, the deep waters of the oceans could begin to release massive amounts of CO_2, driving atmospheric CO_2 higher and faster than other natural processes can reduce it, with still more global warming resulting in an acceleration of the release of carbon dioxide from ocean water and acting in a gigantic and catastrophic positive feedback loop. Counterbalancing this increase in atmospheric carbon dioxide, higher planetary temperatures would speed up the rate of carbonate mineralization of CO_2. But sequestering CO_2 in the form of carbonate rocks occurs slowly, requiring thousands of years, and meanwhile, orbital, tilt, and wobble astronomical effects march inexorably toward the slow onset of another planetary ice age.

What will be the final outcome of the struggle between these titanic climate forces? It is impossible to predict with certainty, but the extremely long duration of the ice-age cycle argues forcefully that ice ages ultimately win out against human-induced global warming, eventually producing gigantic glaciers and ice sheets that endure until the warmth from the sun returns in full measure to the polar regions of the Earth and the massive ice sheets begin to recede once more. It is far from clear how all these factors will play out, but something will certainly happen to Earth's climate, and this something could be very bad indeed.

More predictable, and perhaps more understandable, might be the energy wars fought as the supply of planetary fossil fuels fail to meet global demand. If ruinous civilization-altering wars do occur, the first one might be precipitated by insufficient world oil supply. We are already nearing the point at which the annual demand for petroleum will exceed the ability of the Earth to supply this fuel. Meanwhile, the world's population continues to grow, and climate change is now in progress. The steady growth in the world's demand for energy makes the development of the technologies outlined in this book increasingly important. These technologies, and perhaps others as well, offer approaches for supplying the energy necessary to sustain advanced civilization and the lives of those who dwell within it. Without the development of these technologies, economic upheaval is the very least that can be expected, even without the specter of energy wars. With the advent of major wars over dwindling fossil fuels, food, or other resources, there are no limits to the chaos that would ensue. The Epilogue presents starkly divergent choices that humanity must face.

Epilogue: ORBITuary

ORBITuary. A strange new word, but one signifying something portentous happening to our planet. Comprised of *orbit* and *obituary*, it implies the slow death of something in motion, and that something might be the technological civilization that now exists on the third planet orbiting the sun, a civilization that someday might have built starships. The exhaustion of fossil fuels and the climate changes that could overwhelm our civilization compel us to deal with these difficulties. Nature is unforgiving, to men as well as dinosaurs. Collisions with fast-moving interplanetary objects such as comets or meteors could happen anytime. But such events don't seem as directly threatening as fossil fuel exhaustion, massive increases in atmospheric CO_2 levels, and unprecedented changes in Earth's climate. The British-born American artist and preeminent member of the Hudson River School of landscape painting, Thomas Cole (1801–1848), aptly caught the flavor of a civilization's decay in his painting *Desolation*, reproduced in black-and-white in the image below.

Thomas Cole, *The Course of Empire: Desolation* (1836). Oil on canvas, 19¼ in. × 63¼ in. Collection of the New York Historical Society. Inventory Number 1858.5.

Energy Demand and Climate Change: Issues and Resolutions. Franklin Hadley Cocks
© 2009 WILEY-VCH Verlag GmbH & Co. KGaA, Weinheim
ISBN: 978-3-527-32446-0

Desolation is the last of five paintings commissioned in 1832 to show the development and eventual collapse of an empire. It depicts powerfully what can happen to the mightiest of cultures. Energy might be the Achilles heel of modern civilizations. But the end of the age of energy is not inevitable.

As we have seen, the resources of the Earth include nuclear as well as fossil fuels, and the energy from these nuclear fuels, for all the radioactive waste problems they bring with them, might supply us with energy far longer than the few centuries we have had from fossil fuels. In addition to nuclear energy are those renewable resources from wind, waves, tides, flowing water, and, above all, sunlight. But even if all of these are fully developed, ethnic and religious strife are still loose in the world. Education of both men and women worldwide, the general availability of effective birth control, and better ways of controlling international conflicts all have vital roles to play.

We live in a pivotal time in human history, poised between feast and famine, paradise and hell. Wrong decisions lead to lost time that cannot be found again. Time will tell whether humanity will triumph or self-destruct. If our civilization does collapse, perhaps we but share the fate of other advanced civilizations that may have existed for awhile elsewhere in the galaxy, thereby demonstrating one answer to Fermi's paradox. In our case, all is not yet lost. Technology and science are still advancing, Earth's resources are not exhausted, and Armageddon has not happened. We still have the gift of hope, the gift that never escaped from Pandora's box. In the end, even boundless energy supplies, absent the will to use them wisely, cannot save us from ourselves. But technology, engineering, and science could provide us more time beyond the 10,000 years we've already had since the end of the last ice age. On the scale of millennia, there are only a few thousand years left before Earth's orbital motions again diminish the sunlight received during summertime near the poles and another ice age begins. With fossil fuels gone, these future millennia might be difficult ones indeed.

Humanity must expand non-fossil energy sources until they supply all the energy that we need. There is no other way to maintain an advanced civilization over thousands of years. We must begin working on these new technologies now. Otherwise we might have to get through the next ice age as our ancestors did the last one, huddled around campfires and heaping on more wood as the wind grows chill. On the other hand, with wisdom, hard work, and good fortune, it might be humans who reach the stars.

Credits

1. The data in Figure 2.1 and the photograph in Figure 2.2 are courtesy of the National Aeronautics and Space Administration.

2. The data in Figure 3.1 are courtesy of the National Oceanic and Atmospheric Administration.

3. Figure 3.2 represents a composite set of data from numerous ice core studies made by European, North American, and Russian investigators. The most recent results, extending these data back 800,000 years, are from the European Project for Ice Coring in Antarctica, and their composite data set was used to prepare this figure.

4. The quotation that opens Chapter 4 is used by permission of the University of Washington Press.

5. The photograph in Figure 7.1B was taken by Leslie Deak and is used by permission.

6. The photograph in Figure 7.1C was taken by Elijah Cocks and is used by permission.

5. Figures 7.2 and 7.3 are from the report entitled "Ocean Engineering in Korea" and are used by permission of Dr. K. S. Lee of the Korean Ocean Research and Development Institute.

6. Figure 8.1A is courtesy of the Bureau of Land Management.

7. Figure 8.2A is used by permission of Terrasimco, Inc.

8. The quotation that opens Chapter 9 is used by permission of the Literary Estate of Arthur Conan Doyle.

9. Figure 14.1 is courtesy of the International Thermonuclear Energy Reactor group.

10. Figures 21.1 and 21.2 are courtesy of the National Aeronautics and Space Administration.

Energy Demand and Climate Change: Issues and Resolutions. Franklin Hadley Cocks
© 2009 WILEY-VCH Verlag GmbH & Co. KGaA, Weinheim
ISBN: 978-3-527-32446-0

11. The quotation that opens Chapter 22 is used by permission of A. P. Watt Ltd., on behalf of the Literary Executors of the Estate of H. G. Wells.

12. The Thomas Cole painting in the Epilogue is reproduced in black and white by permission of The New York Historical Society.

Appendix I

Units for Energy and Power

Energy and power are not the same thing. Power is the rate at which energy is used per unit of time. Energy is the ability of a physical system to do work, and the word energy itself is derived from the Greek word meaning operation or activity. The units for power and energy are related by units of time. One joule of energy is enough to lift 9.8 kilograms of matter one meter in the Earth's gravity field. The factor of 1/9.8 appears here because one kilogram of matter on Earth, subject to its gravitational attraction, will accelerate 9.8 meters per second squared when it is dropped. For comparison, on the moon one joule is enough energy to lift 1/1.6 kilograms of matter one meter, because the moon's gravitational attraction will accelerate it only about 1.6 meters per second squared when it is dropped. One watt of power, on the other hand, is the use of one joule of energy per second. So, continuously lifting 1/9.8 kilograms of matter on Earth one meter upward every second requires one watt of power.

These fundamental definitions of energy and power are not commonly used. But everyone knows that it takes energy to boil water, and the boiling of water involves heat. One gram of water (which at room temperature is also extremely close to being one cubic centimeter of water) can be heated one degree centigrade by 4.184 joules of energy, and this amount of energy is defined as one scientific calorie, which isn't all that much. One scientific calorie equals the energy produced by 4.184 watts operating for one second, and a watt-hour of energy equals 15,062.4 joules. A food calorie is defined as a thousand scientific calories, or 4184 joules. An older unit, still very much in use in North America, is the British thermal unit (Btu). One Btu is the heat necessary to increase the temperature of one pound of water one degree Fahrenheit. One Btu equals 252.16 calories. Heat is a form of energy, and energy in the form of work can be completely converted into heat. But the opposite is not true. Heat cannot be converted completely into work for fundamental thermodynamic reasons.

One thousand watts of power equal one kilowatt, and one million watts are a megawatt. A large coal-fired power plant produces 1000 megawatts of electric power, thereby delivering the amount of energy needed to lift 1,000,000/9.8 kilograms of matter one meter per second or to heat about 239,000 kilograms of water one

Energy Demand and Climate Change: Issues and Resolutions. Franklin Hadley Cocks
© 2009 WILEY-VCH Verlag GmbH & Co. KGaA, Weinheim
ISBN: 978-3-527-32446-0

degree centigrade per second, or about one million pounds of water one degree Fahrenheit per second. That's a lot of power.

From knowledge of various prefixes, one can unravel many of the other units used to describe power and energy, such as terawatts (one thousand megawatts) or exajoules (1,000,000,000,000,000,000 or 10^{18} joules). An exajoule of energy would be produced by a 1000-megawatt power plant operating for 31.7 years. The world currently uses over 400 exajoules of energy per year. There are so many other prefixes for denoting energy and power magnitudes that they can become confusing. That's why this book has tried to stay with watts and megawatts and joules and megajoules.

One older term for power still in common use is horsepower. It seems that a strong horse can lift 550 pounds one foot in one second, which translates to the horse lifting 249.7 kilograms 0.305 meters in one second, under standard gravity conditions. That is one horsepower, or 42.4 British thermal units (Btu) per minute, or 746 watts, but no one uses Btu per minute or watts to rate the power of engines in vehicles. A Btu is the amount of energy it takes to heat one pound of water 1°F.

But a note of caution: pounds are sometimes used as units of force and sometimes as units of mass. Mass and force are not the same. For example, a pound-mass weighs one pound force under standard gravity conditions, and a kilogram mass weighs 9.8 newtons, but the weight of a pound-mass or a kilogram of matter on the moon is much less. Fortunately, the power usage of lights, appliances, and electrical devices in general is usually given in watts. A toaster uses about 1000 watts of power, which means that it needs more than one horsepower to make toast, but horsepower is never used to measure the rate of energy used by toasters. Although very few of us think of toasters in this way, it is nonetheless true that toasters are about one-horsepower devices, just as cars are about 100-horsepower devices. Different units for energy and power are appropriate for different applications, but joules, calories, watts, and horsepower are energy and power units that are used, for one purpose of another, almost everywhere.

Appendix II

Units for Radiation

Measurements of radiation can get confusing. A *roentgen*, one of the first units used to measure radiation levels, is proportional to the quantity of energy absorbed by dry air from the radiation passing through it. This unit is named after the German physicist Wilhelm Roentgen (1845–1923), who received the first Nobel Prize for physics in 1901 for discovering x-rays. People are not made of dry air, and the radiation absorbed by tissue or other substances is measured in units named after the British physicist Louis Harold Gray (1905–1965). A *gray* is defined as one joule of energy absorbed from incident radiation by one kilogram (2.2 pounds) of the material irradiated, whether this material is tissue, air, or anything else. One calorie (4.184 joules) of energy will heat one gram of water one degree centigrade. For flesh, one gray is about 100 roentgens, but this relationship varies with the type and energy of the radiation. As absorbed, different types of radiation can have different biological effects. To account for this and confuse matters even more, radiation dosage in grays is adjusted by a quality factor for the varying bio-logical effects of different types of radiation. High-energy particle radiation, such as protons and electrons, tends to do more biological damage than high-energy electromagnetic radiation such as x-rays and gamma rays.

To take this difference into account, the absorbed dose in grays is multiplied by a so-called quality factor to give the radiation dose in *sieverts*, named after the Swedish medical physicist Rolf Sievert (1896–1966), who developed methods for standardizing the radiation doses delivered during cancer therapy by the differing radiation techniques used in various hospitals.

Ionizing radiation has sufficient energy to break atomic bonds and produce ions, that is, atoms that are missing at least one electron. Radiation therapy requires ionizing radiation. A sievert represents a large amount of ionizing radiation, and millisieverts are usually used to describe radiation dosage. For example, a normal chest x-ray delivers about 0.1 millisieverts of radiation to chest tissue. A whole-body dose of 5000 millisieverts is usually fatal.

An older unit that preceded the use of grays is the *rem* (short for roentgen equivalent man), and this unit is still in use. One sievert equals 100 rem. Another measure of radiation is the *curie*, named after the Polish-French physicist Marie

Energy Demand and Climate Change: Issues and Resolutions. Franklin Hadley Cocks
© 2009 WILEY-VCH Verlag GmbH & Co. KGaA, Weinheim
ISBN: 978-3-527-32446-0

Sklodowska Curie (1867–1934), the first woman to teach at the Sorbonne, and her husband Pierre. With Antoine Becquerel, they shared the 1903 Nobel Prize in physics for the discovery of radioactivity. She and Pierre also discovered the elements radium and polonium, which she named in honor of her homeland. In 1911 she was awarded the Nobel Prize in chemistry for isolating pure radium. The original definition of a curie was the disintegration rate of one gram of radium, but it is now defined as 37 billion radioactive disintegrations per second. Radioactive isotopes are usually sold in units of millicuries or microcuries rather than grams or ounces.

Index

Energy Demand and Climate Change: Issues and Resolutions. Franklin Hadley Cocks
© 2009 WILEY-VCH Verlag GmbH & Co. KGaA, Weinheim
ISBN: 978-3-527-32446-0